Aviation and Human Factors

How to Incorporate Human Factors into the Field

Aviation and Human Factors

How to Incorporate Human Factors into the Field

José Sánchez-Alarcos

CRC Press is an imprint of the
Taylor & Francis Group, an informa business

CRC Press
Taylor & Francis Group
6000 Broken Sound Parkway NW, Suite 300
Boca Raton, FL 33487-2742

First issued in paperback 2023

ISBN 13: 978-1-03-257105-8 (pbk)
ISBN 13: 978-0-367-24573-3 (hbk)

DOI: 10.1201/9780429283246

Library of Congress Cataloging-in-Publication Data

Names: Sánchez-Alarcos, José, 1957- author.
Title: Aviation and human factors : how to incorporate human factors into the field / by José Sánchez-Alarcos.
Description: Boca Raton, FL : CRC Press/Taylor & Francis Group, 2020. | Includes bibliographical references.
Identifiers: LCCN 2019008023 | ISBN 9780367245733 (hardback : acid-free paper) | ISBN 9780429283246 (ebook)
Subjects: LCSH: Aeronautics--Human factors.
Classification: LCC TL553.6 .B35 2020 | DDC 629.13--dc23
LC record available at https://lccn.loc.gov/2019008023

Publisher's Note
The publisher has gone to great lengths to ensure the quality of this reprint but points out that some imperfections in the original copies may be apparent.

Visit the Taylor & Francis Web site at
http://www.taylorandfrancis.com

and the CRC Press Web site at
http://www.crcpress.com

A los que permanecen con nosotros y al que nos dejó.

Contents

Preface

This book has a story to be told: It was intended to become the second edition of *Improving Air Safety through Organizational Learning*, published in 2007, with updated information. However, many important things have happened since that publication. An update is not enough to reflect what has been going on and what are the new problems and possible new solutions.

In some ways, the previous book anticipated the dangers related to organizational evolution based on technology while disregarding other options. The following text, taken from the 2007 publication, was pointing to something that would happen only a few months later.

Often, automatisms act by default, without knowledge of the operator, appearing, when faced with unforeseen events, to restrict action rather than to assist it.

Two accidents that could be described using the paragraph above, XL888T and AF447, took place in a very short time. Crewmembers were confused about what was happening, and misunderstood automation had a key role in the outcome. More recent accidents, like those related with B737MAX, are still better described by the statement.

XL888T and AF447 accidents shared some important features: In both cases, a failed sensor caused the plane to behave in a bizarre way without the pilots knowing what was happening. No further technical failures resulted from the frozen sensor: The different systems of the plane behaved as expected from their design. A single wrong sensor was enough to trigger the disaster.

It seems that a new model of accident was born, driven by confusion about what the plane was doing and why. Why did it happen and how can it be avoided in the future?

Some people would suggest that there is nothing to worry about, since safety rates keep improving, albeit at a slow pace. Risk, as everybody knows, is a two-faced coin composed of severity and probability, and it seems that the whole system has been working on the basis of probability at the cost of increasing severity.

The whole risk account can appear favorable, but how many common-mode situations could be buried under millions of program lines or produced by an explosion that severs data wires affecting different systems?

Could a faulty sensor – as with AF447 and XL888T – again trigger a state of confusion in the flight crew? Could we find a problem affecting thousands of planes sharing the same systems or the same manufacturing or maintenance processes?

Are the systems in a plane fully protected against unauthorized access? We should not forget that some manufacturers and regulators are speaking openly about single-pilot planes that can be controlled from the ground. Is a Zero-Day attack, affecting many flying planes at the same time, unthinkable?

To paraphrase Nassim Taleb, we might ask ourselves whether we are running a "black swan" farm.

Please do not read the preceding paragraphs as an indictment of technology. That would be crazy, especially in a field that could not exist without technology. The position you will find in this book is very easy to explain.

1. Not everything that is technically feasible is, at the same time, advisable.
2. Trading understandability for efficiency is almost always wrong. It is especially wrong if we can afford a certain amount of inefficiency (e.g. IT design) for the sake of understandability.
3. Aviation does not have the privilege enjoyed by other fields (e.g. NPPs or railways) of freezing everything while we look for the solution to the problem. The solution must be found while the plane is flying and with the limited resources that can be found on board and with a helping hand from ATCs.

Aviation, as it is right now, is an example of Perrow's classic *tightly coupled systems* definition. Those systems – designed to be, above all, efficient (i.e. having an optimum input–output rate) – can suffer snowball effects coming from that tightly coupled process. Therefore, an efficient organization is not only efficient during normal operation; it is also efficient in the outcomes of its errors.

So, if we want a serious improvement, we should decouple these systems by creating firewalls and by looking for real understandability.

An operating model that is not understood by the operator will be adequate only while the system is working as expected. Once the system shows a bug or launches the wrong automatic process, the operator is confronted with a situation where the operating knowledge is useless.

In a YouTube video published by Airbus, a cockpit is shown, and the pilot says that we do not need to know how a computer works to operate it. To emphasize his position, he holds up a pocket calculator, telling us that we do not know how it works and, despite that, we can operate it. Unfortunately, that is not true.

We know how a calculator works because it is designed to imitate us, even if that means decreasing efficiency in the operation. The RPN calculators manufactured by Hewlett Packard are an interesting example: These calculators are loved by engineers and finance specialists but difficult to understand for many people. The most remarkable feature is the absence of the "=" key, making things difficult for novice users. After years selling RPNs, HP had to manufacture less efficient arithmetic calculators; otherwise, they would have given up the biggest chunk of the market. Amazingly, many RPN calculators designed and manufactured for the first time more than

30 years ago are still sold and used, an uncommon record for an electronic device.

Unlike pocket calculators, in Aviation the "market" is not composed of hundreds of thousands of individual pilots, controllers, technicians or any other people involved in the operation and maintenance of planes.

Of course, all of them are stakeholders, since they put their jobs and, in some cases, their lives on the line. However, judging by the effects analyzed here, the market is a very selective club composed of manufacturers, operators and regulators immersed in a market-driven dynamic that could be heading down the wrong path.

- Operators ask for efficiency everywhere – that is, in operating the plane, in maintaining it and in the training required for both things.
- Manufacturers compete to meet these requirements by including more electronics, cheaper manufacturing techniques and commonality in their planes.
- Regulators know that the aviation industry is important, and they do not want to risk killing or damaging the market by raising their own requirements more than necessary. Instead, they can be aligned with their reference manufacturers, especially the big ones.

The figures on air safety are good – albeit not everywhere – but despite these figures I would like to ask a single question:

Is it acceptable that a plane can fall from 10 km above the ground with its engines and its controls working perfectly, while the pilots know that they are going to crash but not why? Is that a matter of training or is it a deeper issue with its roots in an organizational model?

Different cases will show that *lack of training* could be but one piece of a very complex organizational puzzle. That's why a change of course would be advisable. This book deals with that situation.

Author

José Sánchez-Alarcos has a PhD in sociology and a BS in psychology. He is a pilot and the author of *Improving Air Safety through Organization Learning.* He is highly experienced in the human and organizational factors behind air safety. He has served as a Human Factors trainer at EASA and SENASA, and for many different manufacturers, airlines and aviation authorities. As a Human Factors consultant, he has worked in Aviation design and accident investigation in Aviation, nuclear installations, railroads, and maritime transportation.

List of Abbreviations

AI	artificial intelligence
APU	auxiliary power unit
ASRS	Aviation Safety Reporting System
ATC	air traffic control
CAA	Civil Aviation Authority
CAWS	central advisory and warning system
CRM	crew resource management
CVR	cockpit voice recorder
EASA	European Aviation Safety Agency
EID	ecological interface design
ETOPS	Extended-Range Twin-Engine Operation Performance Standards
EWIS	electrical wiring interconnection system
FAA	Federal Aviation Administration
FBW	fly-by-wire
FDM	flight data monitoring
FDR	flight data recorder
FRMS	fatigue risk management system
GOFAI	good old-fashioned artificial intelligence
HFACS	Human Factors Analysis and Classification System
IATA	International Air Transport Association
ICAO	International Civil Aviation Organization
IFALPA	International Federation of Airline Pilots
ILS	instrumental landing system
MEL	minimum equipment list
NASA	National Aeronautics and Space Administration
NOTAM	notice to airmen
NTSB	National Transportation Safety Board
RAAS	runway awareness and advisory system
RPN	reverse Polish notation
RVSM	reduced vertical separation medium
SA	situation awareness
SOP	standard operating procedure
SRK	skills, rules and knowledge
TEM	threat and error management
TOWS	take-off warning system
VNAV	vertical navigation

1

Commercial Aviation: A General Picture

Compared with many other activities, Commercial Aviation is still relatively young. However, the relevance – in terms of the number of users – as well as the improvement in safety levels are impressive.

It is far from being a failed model, but pressure to improve grows with the number of flights and with the size of the planes, increasing the potential impact of a single event.

As could be expected, it is a highly technical field, but some of the potential problems come precisely from disregarding the non-technical aspects, both on the individual side and on the social side.

This first chapter is intended to be a warning sign. In the following chapters, this warning will be developed through different aspects, showing the organizational evolution, its flaws, what should be done to include – seriously – the human side in the future and why this is a must.

Aviation: A High-Stakes Field

Organizational learning is a major issue in many fields. Some mistakes are repeated time and again, making it clear that learning does not exist or is a very slow process.

Regarding this issue, Aviation figures define it as a successful case, especially since it is a young activity, compared with many others.

Little more than a century has passed since the first powered flight, and it would be hard to imagine the world today without millions of people flying everywhere every single day.

The last available statistical report from Boeing gives us an idea about the relevance of the activity as measured by recent growth (Figure 1.1).

The advances in safety since the early days of Commercial Aviation are still more impressive than the amount of activity. Actually, these advances can be used as an index to measure the success of Aviation in terms of its learning ability.

Aviation comes from a situation where the ordinary retirement of a pilot was a rarity, while nowadays that is the expected outcome. Experienced passengers, even average ones, do not think more about the chances of an accident when they board a plane than when they board a train or a bus. Data on accident rates, from the same source, are shown in Figure 1.2.

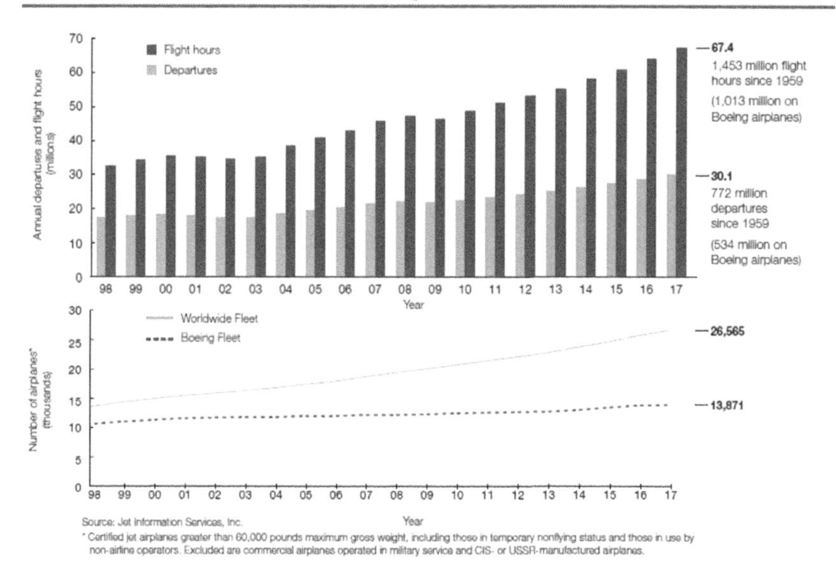

FIGURE 1.1
Departures, flight hours and jet airplanes in service. (From © 2018 The Boeing Company. All Rights Reserved.)

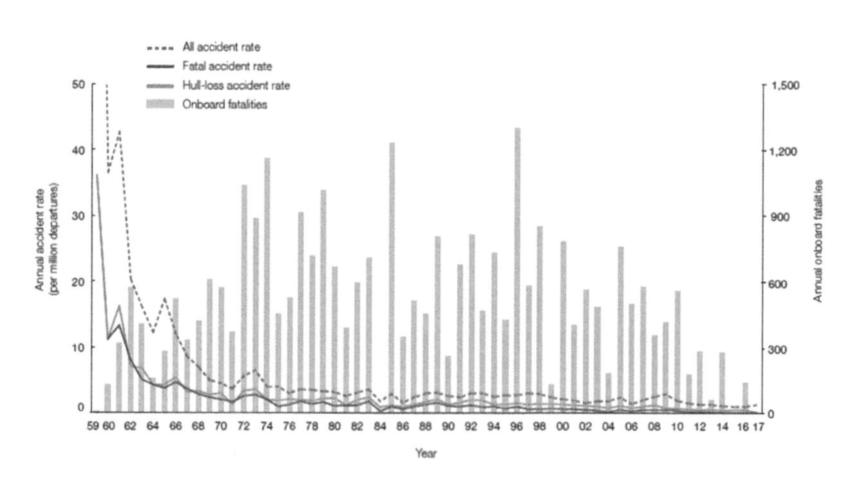

FIGURE 1.2
Accident rates and on-board fatalities by year. (From © 2018 The Boeing Company. All Rights Reserved.)

The success is greater still given Aviation is a field where major dangers remain or even have grown over time: A high-level view inside a single commercial flight is enough to appreciate the real value of the current safety requirements. That view is represented in the Decalogue of Major Flight Dangers.

DECALOGUE OF MAJOR FLIGHT DANGERS

1. Commercial jets, at their cruise altitude, move near to the speed of sound. Any impact against another object, whether it is moving or not, means disaster, and the time available to react is very limited.
2. Aircraft travel about ten kilometers above the ground. The atmosphere is not breathable, and to the risk of asphyxia must be added the risk of impact due to uncontrolled loss of altitude.
3. The external temperature at cruise level is around −50°C. Exposure at these temperatures would not permit survival beyond a few minutes.
4. External atmospheric pressure is very low. To maintain a comfortable pressure inside the aircraft, the sheet metal that separates the interior from the exterior holds significant pressure, producing metal fatigue and risk of explosion.
5. On long-haul flights, half the take-off weight of the aircraft can be fuel. In the case of impact, this involves a risk of fire or explosion, and in the case of depletion, grave risk of accident.
6. Engines work under high-pressure and high-temperature conditions, with the associated risk of mechanical failure, explosion or fire.
7. Large aircraft, under almost any weather condition, take off at speeds close to 300 km/h and land at speeds close to 200 km/h.
8. Aircraft are subject to all sorts of weather that can affect visibility and cause structural impacts, electrical shocks or build-ups of ice on their external surfaces.
9. Aircraft cover long distances over all types of geographical zones. At some geographic points, a twin aircraft could be almost 6 flight hours away from the nearest airport.
10. Congestion on some flight routes or in the terminal zones of large airports increases the risk of collision.

The list could be far longer, but this is not the most important point. The really important issue here is the difference between the concepts of *danger* and *risk*.

While danger is related to severity (i.e., a list of increasingly severe dangers) risk includes probability. The difference between these two concepts, which for many lay people are synonymous, is reflected in the decreasing accident rate. In other words, while danger remains or grows, and nothing can be done about it, risk decreases and, hence, the accident rate decreases too.

A good example is the Airbus A340 model: When Airbus, after 386 units manufactured over 18 years, closed the assembly line, no one had been killed in a major accident on the millions of flights completed.

This might encapsulate the relevance of safety improvements in Aviation but, of course, there are other activities that could compete successfully, at least in terms of relevance.

Private transportation – "the machine that changed the world", as the car was called – or the Information Technology revolution have had a larger impact still. However, Aviation has an important feature: the enormous impact of a single major accident.

Car accidents are common, and the death toll on roads is far higher than in Aviation. Actually, the death toll in the United States alone in 2017 was an estimated 40,000 people, enough to fill 100 jumbo jets. However, every single accident has a minor impact, compared with a major Aviation accident. Since every car accident has a small number of victims, acceptability is far higher for 1,000 victims in 1,000 different accidents in a year than for the same number in three accidents in the same period.

Wells (2001) calculated the public impact of an accident considering its direct relation to the number of victims, such that the loss perceived by society is proportional to the square of the number of deaths in a single accident; so, 100 deaths in one accident cause the same social impact as 10,000 deaths produced in individual accidents. Beyond the calculations of the social impact, anyone is able to imagine what would happen if 100 jumbo jets crashed in a single year in the United States (to relate car and Aviation figures to the same area).

This perception has a practical side: Aviation is always under public scrutiny. Therefore, pressure to keep or increase safety level is higher than in other activities. However, some other activities bring their own dangers and, since a good part of Aviation advancement comes from them, Aviation has imported these dangers from outside.

The best example is Information Technology: Cybersecurity is still a new issue even though, right now, it is a high-stakes activity, since basic installations depend of it. Of course, that includes Aviation-related installations.

The impact of Information Technology is such that an electromagnetic pulse (EMP) could send the affected area back to the Stone Age, but that is not the only threat: Guided-software attacks against specific installations could be very damaging and even more costly in terms of human lives than many Aviation accidents.

Cyberattacks are not the only danger: Poor design or software bugs can have important consequences. It is not a matter of comparing the relevance of Aviation and Information Technology but highlighting a single point: Information Technology is right now a key part of Aviation and, as such, Aviation has imported some dangers coming from a ubiquitous Information Technology:

avionics, control through fly-by-wire, datalink communication, air traffic control panels, onboard anti-collision systems, instrumental landing systems, onboard and ground installations, next-gen traffic control systems – it is hard nowadays to find a single system or activity in Aviation where Information Technology is absent or has a secondary role.

Planes such as the Boeing 787 or A350 use software involving millions of lines of code. The amount is far below that of a modern fighter jet, but still a good place for hidden bugs or attacks.

However, intentional damage is not yet the biggest danger coming from Information Technology. The biggest danger comes not from mistakes during the design process or from the activity of hackers, on their own or sent by an external intelligence service. The biggest and most frequent danger – that is, the biggest risk – is related to software performing its task as designed but in the wrong context.

Leveson (2004) addressed this problem, warning about the confusion between safety and reliability, pointing out that "in complex systems, accidents often result from the interaction among components that are all satisfying their individual requirements, that is, they have not failed".

US1549 is a good example. The manufacturer, Airbus, never foresaw a full loss of power below 20,000 ft. Every procedure and every system worked perfectly – for a plane flying at 20,000 ft. or higher. Had the situation been managed by an advanced system, instead of humans, designed under the same parameters, a full hull loss and a significant loss of lives could occur, yet everyone could say that the whole system had worked properly – that is, according to the design parameters.

Information Technology has brought new facts into the fold, and both safety and efficiency have been a major issue. However, safety and efficiency don't work well together. Different authors have shown that the safety level must balance protection and operating costs. Losing that balance can drive bankruptcy – or disaster and probable bankruptcy.

Flying implies risk, and risk must be kept within acceptable levels. Once there, the safety–efficiency tradeoff is an acceptable practice.

However, nobody can permanently define the borders of acceptability and, hence, nobody can say where these borders will be tomorrow. The borders will always be by-products of public opinion, largely influenced by how recently a major accident has occurred, and how and why.

Then, safety levels become a moving target through acceptability and its evolution. Acceptability, as measured through public opinion, could be a time bomb if disregarded in advance, directly or by assigning a probability and forgetting anything below it.

Anyone can decide on a transaction between efficiency and safety as long as "an appropriate level of control" and "the appropriate balance between production and protection" are guaranteed. However, if a major event shows that a hidden transaction took place, the acceptability level can change instantly and the speed at which information spreads – also a feature of the Information Technology age – can transform a local issue into a market earthquake. The B737MAX issue could be a good sample.

Risk, as expected in a highly technical field, is acceptable or not depending on the trust that users give the specialists instead of their own evaluation ability. However, as Beck (1992) points out, it is not a static situation. Trust in specialists can be lost with catastrophic effects.

> The evidence of the collateral effects of products or industrial processes that are putting at risk the basic requirements for life could drive to market collapse, destroying political trust, economic capital and *the belief in the superior rationality of the experts.*

In the past, with an information flow that was easier to manage, airlines have disappeared after a major accident. Furthermore, the disappearance of the big manufacturer McDonnell Douglas can be traced back to its unfortunate DC10 model.

> The TK981 accident revealed bad practices by both the manufacturer and the regulator at conspicuous levels. That would lead to a stampede of potential passengers who did not want to fly in that plane and an overreaction from officials trying to put up a firewall against the situation.

All operators are now conscious of a new business risk: Anyone can communicate anything to the whole world. If that "anyone" happens to be an *influencer,* any wrong action can become a *trending topic* in a matter of hours. Community managers – a new position that did not exist a few years ago – can try to prevent the impact, using all the resources to hand, but it could be unstoppable.

In the past, technical decisions were decided in forums, and it was a comfortable position – comfortable enough to allow misbehaviors such as the DC10 example. Nowadays, something like that would be known about in a much shorter time. Furthermore, a decision based on statistics, if hidden to a public that would see it as inadequate, could have the same results.

> The first known fatal accident with a Tesla car driven by autopilot produced an awkward statement from the company, who claimed on its website that autopilots are safer than human drivers. That position irritated many people more than the accident itself.

Safety is not only a matter of statistics. It is a matter of public acceptance and, hence, it is a matter of transparent behavior. Otherwise, the comfort of decisions made only by specialists could become a business nightmare with invaluable consequences.

Many examples can be found of how the public dimension is handled. The aforementioned Tesla example is clearly among them, but after a severe penalty from the Federal Aviation Administration (FAA), Southwest Airlines took a very different approach: They explained on their website why they did not share FAA's view and subsequently uploaded CEO Herb Kelleher's testimony before the judge.

In short, public scrutiny cannot be bypassed. A situation could be acceptable from a statistical point of view but unacceptable for users. This is not only a matter of the rationality of the decisions: When an unlikely situation occurs, so unlikely that its probability is supposed to be below one in a billion, the investigation usually reveals that the estimation of the probability had been unjustifiably low.

> One of the clearest cases of this situation is the United-232 accident. The chance of losing three redundant hydraulic systems that worked perfectly at the beginning of the flight was supposed to be below one in a billion. The accident would show that, despite the three systems being functionally independent, they were in very close proximity in some parts of the plane. A single event – an explosion of the tail engine in this case – could affect all of them. The event was enough to show that the estimation of the probabilities was entirely unrealistic.

Different situations demonstrating this feature will be discussed and, in almost all of them, a single issue will emerge: the decreasing ability of people to detect, manage and correct problems stemming from a erroneous technology-based development.

People, especially people close to the operation such as pilots or controllers, are seen very often as emergency resources but, to work properly as such resources, some conditions must be met. Otherwise, Aviation will confront both the operational and the public perception risk. Statistics show a bright present, but statistics are not enough, especially those based in a world as planned, not as it really is.

The Fight for Efficiency

In an environment subject to the threat of sudden changes in the acceptability of safety levels, a hard fight for efficiency must be fought. An analysis of the evolution of the two major manufacturers, Airbus and Boeing, will show how efficiency always won the fight against safety.

When Airbus introduced the A310 in 1978, it came with a new concept: the forward-facing crew cockpit (FFCC). It was the first time that a large passenger transportation plane had removed the flight engineer from the cockpit.

Boeing complained, but B757 and B767, then in the design process, appeared in 1981 without a place for the flight engineer. In 1998, a major accident occurred that could have invited a review of this practice, but many years had passed since it had become an operating standard.

> SW111, after a fire on board, tried to land in an unknown airport. Two people had to fly the plane, navigate the landing, communicate and, in their free time, dump fuel and try to fix, if possible, the problem that provoked the fire.
>
> Nobody could affirm, by any means, that the presence of a flight engineer would have saved the plane, but the description of the situation suggests that another pair of smart hands could have helped. However, the plane, as designed, did not have anywhere for them to work.
>
> Interestingly, the conclusions of the official report do not have a single mention of the word "workload". There is only a vague reference to it while speaking about there being maps in the wrong place, adding a task to the situation.
>
> The accident happened because of incorrect installation and lack of training – an old friend that can be used time and again as a good one-size-fits-all explanation.

Hence, nobody suggested the idea of reverting – or, at least, reconsidering – the practice that removed the flight engineer from the cockpit years before.

Boeing had its own turn in the search for efficiency. They had seen that engine stops were uncommon in long flights inside the United States. Then, they suggested that engine reliability could justify the use of twin-engine planes to perform oceanic flights.

Of course, Airbus complained at the first opportunity, since this suggestion directly competed with its A340, a four-engine plane designed for long routes with fewer passengers than a B747. Therefore, flights performed by A340s could be made by twin planes, smaller than B747s, if the proposal was accepted. That would lead to the birth of the Extended-range Twin-engine Operational Performance Standards (ETOPS), which will be widely commented on.

Models with three engines such as DC10 from McDonnell Douglas or MD11 from Lockheed Tristar would sit in the middle without the advantages of four or two engines.

In the meantime, the authorities were cautious. The collection of information about engine reliability drove an increase, in successive steps, in the allowed time out of an airport – recently more than 6 hours. Airbus, at the beginning, complained, but it launched the twin A330, addressed precisely to the long-range market, performing oceanic flights.

Right now, four-engine planes are a rarity, only justified by huge planes that cannot fly with only two engines. They would be so massive that – keeping the usual design – they could not be installed under the wings without dragging them or making landing gear legs so long that their resistance would suffer.

The present ETOPS rules accept that a twin plane, after an engine stop, could continue for more than six hours before reaching an airport. In other words, a plane loaded with more than 300 passengers could legally fly without a suitable place to land, with only one engine, for more than six hours.

The statistics say that this is a sound decision, but passengers' guts could feel otherwise. Manufacturers and designers try to fight any common-mode situation that could be imagined. For instance, there are maintenance rules preventing the same teams from intervening in both engines at the same time. However, the threat of a common mode still exists, including the fact that the large amount of data for statistical analysis come from two engines working as they should.

Under this statistically supported practice, there is a practical issue: Nowadays, twin planes can fly everywhere with no more limits than a four-engine plane. Hence, the old advantage of four-engine planes disappeared since the routes previously forbidden for twin planes are already allowed.

Therefore, four-engine planes will exist only for as long as engineers cannot put all the power needed to lift a huge plane (e.g. an A380 or a B747) inside a little engine without breaching ETOPS acceptability in the event of a genuine *black swan* as defined by Taleb – that is, an unforeseen event decreed as unthinkable.

How could that happen? Statistics speak a lot about reliability levels. Then, should we – as our guts could suggest – be worried about the possibility of overflying an ocean with a single engine for more than six hours if our plane suffers an engine stop in the worst imaginable place?

Engines are exceptionally reliable, but do they have the same reliability level in all operating conditions? For instance, once an engine fails, the remaining engine must supply all the power needed to keep the plane flying. Is there enough information to guarantee that the reliability level is still at the same level as when both engines are working?

Curiously, the question is not as easy to answer as expected. An engine stop is an extremely uncommon event and, hence, we could miss information about reliability while flying with a single engine. The bulk data on engine reliability come from flights with both engines working. Then, the reliability itself could be the factor driving us to miss information about reliability in abnormal situations.

Someone could say that we already have that information and, furthermore, twin planes are overpowered.[1] Every flight has to take off and climb, phases in which pilots apply high power settings. The settings for a long flight with only one engine running are higher than usual cruise settings but still lower than required when high power is needed for a brief time. Hence, since every flight has take-off and climbing phases, someone could claim to have all the required information about the reliability of engines working at high power settings.

However, this is the point where things become interesting. Figure 1.3 shows the parameters and the results of a search in the ASRS reporting system database.

Obviously, 292 reported engine stops in the United States in 18 years is an exceptionally small number. However, a sample of these reports will show

Your search returned 292 ACNs

Search Criteria:
Date of Incident was between January-2000, January-2018
and **Federal Aviation Regs** (FAR) Part was Part 121, Part 135
and **Event Type** was Critical, Less Severe
and **Primary Problem** was Aircraft
and **Text** contains engine AND stop

FIGURE 1.3
ASRS search box.

that many of the engine stops happened precisely during take-off or while climbing – that is, the phases where the engines run at higher power.

Even though the number is still low, the safety–efficiency transaction is clearly at an *acceptable level* – as defined by the International Civil Aviation Organization (ICAO). While nothing happens, passengers are subject to a transaction that they are ignorant about, since few of them are familiar with the ETOPS rules.

If the unthinkable – an independent or supposedly independent double engine stop – happens in a wrong place, users, unfamiliar with a transaction that they would not accept, could reject flying twins on oceanic flights. That would provoke major trouble in the system, and some people would have a hard time trying to explain why, despite the event, the decision and the acceptance of the rule was reasonable.

The funny part is that the risk of flying twin planes in such remote places is not a primary risk. First, to confront that risk, someone must decide to overfly places 6 hours away from a suitable airport, frequently loading more than 300 passengers. Once that is decided, an engine stop is not the only possible event and not even the worst.

Whatever the number of engines, emerging ailments that could affect many people on board without more resources than those inside the plane or with very limited support from the ground; depressurization, cabin smoke events, fire, electric failures and many other things can happen in so long a time.

The ETOPS practice, then, is not as unreasonable as it might appear to lay people – once flying in remote places is accepted, instead of using longer routes, closer to ground resources.

However, the problem remains: Instead of providing the public with unsweetened information, a major decision involving risk has been made unbeknownst to the people it affects. This is preferable from the managerial point of view, but it has a serious drawback: An ugly and avoidable side-effect is the risk of a loss of public acceptability if a major event happens.

Of course, the evolution did not finish with the disappearance of the flight engineer and the ETOPS practice. Airbus introduced a new challenge: *fly-by-wire*. In short, fly-by-wire means that the flight controls give orders to a computer instead of having a direct mechanical/hydraulic link with the controlled surfaces.

Before Airbus, the only civil plane with fly-by-wire controls was Concorde. Boeing complained despite fly-by-wire being standard in Military Aviation, where Boeing already had ongoing activity.

Anyway, Boeing finally introduced fly-by-wire in its models B777 and B787 – those where it was possible. Fly-by-wire is a major change and, hence, adapting it for its older models would imply a new certification process.

The development of a new model and its certification is so costly that companies will want to extend its life as far as possible, especially if many planes have been sold and the cost of adaptation to the renewed model is far lower for operators than acquiring a new model . That is why Boeing keeps updating a model from 1967 (i.e. Boeing 737) instead of developing a new one for the short- and medium-range market. Actually, the events related to the B737MAX, whose origin are in the fact that new engines did not have room on the airframe approved in 1967, can give an idea of the relevance of this factor.

Criticizing fly-by-wire was a way for Boeing to defend not only the old B737 but the B747, the B757 and the B767, all of them already on the market when Airbus launched its first fly-by-wire plane. However, when Boeing presented modern designs B777 and B787, they included the formerly criticized fly-by-wire.

If we look at the most modern planes from the two biggest manufacturers, Boeing 787 and Airbus A350, we will find that, of course, none of them has a flight engineer. Both are twins and ETOPS certified and, of course, both are controlled through fly-by-wire systems. Additionally, there are other similarities, such as the amount of composite materials used throughout the whole plane.

Since they are subject to similar constraints, both manufacturers produce similar products where efficiency wins. Then why the initial resistance? The answer is that, as in the fly-by-wire case, they are trying to extend the life of old designs, and the only way is to attack evolution – before accepting it – when it comes from their main competitor. Time and again, facts have shown that criticism coming from the main competitor must be taken with a grain of salt.

However, both manufacturers could find a way around their high development costs. If so, they would not need to hold on to extremely old models as a way to avoid the full certification process.

The eventual solution, something we could see in the coming years, lies in a single word: *similarity.*

We can find in the European Aviation Safety Agency (EASA)'s CS25 specifications this statement about similarity in terms of human error prevention.

> Statement of Similarity (paragraph 6.3.1): A statement of similarity may be used to substantiate that the design has sufficient certification precedent to conclude that the ability of the flight crew to manage errors is not significantly changed. Applicants may also use service experience data to identify errors known to commonly occur for similar crew interfaces or system behaviour. As part of showing compliance, the applicant should identify steps taken in the new design to avoid or mitigate similar errors.

Of course, the concept of similarity does not apply only to the area of human factors. It appears to be related to different systems and it became crucial years ago when Airbus produced its A320 model.

In some ways, A330 was a more grown-up A320, and A340 was an A320 with four engines. The lineage of A318, A319 and A321 is so clear that it does not require comment. Airbus called this practice *commonality* and, based on that, it was possible to have extremely short training times from one Airbus model to another and to keep licenses alive for more than one type through *cross-crew qualification*.

The required time to train a pilot from one model to another, according to the training brochure from the manufacturer itself is extremely short.[2] Going from an A330 to an A340 or the reverse requires 3 working days. These 3 days increase to 15 days from/to an A320 to/from an A380.

Obviously, reducing training time and being able to use the same crews on different planes are new efficiency successes, albeit with some drawbacks, but the advantages could go beyond efficiency in the training and use of crews. They could affect design too.

As said, the keyword is similarity, emphasized as commonality by Airbus. The meaning of similarity in terms of the certification process is this: A system can be certified if the manufacturer proves that it is already working on a plane previously certified and flying.

This practice, used for individual systems or devices, could in a limited way be used for a full plane. If there are many shared configurations with a plane already flying, that could accelerate the entire process for a new model to appear.

Then, manufacturers could produce the same plane in varied sizes and ranges and with minor changes – something that they already do – as versions of the same model, but it could be extended beyond that. We could expect the Boeing B787 to become the Boeing equivalent to the Airbus A320 – the common platform to repeat in varied sizes and ranges – while the Airbus A350 could take the place of the Airbus A320 as the origin of a renewed Airbus family. The solution appears perfect in both dimensions of efficiency: operation and design. But what if something is wrong?

As said, Charles Perrow warned against pursuing efficiency at any cost with the concept of *tightly coupled organizations*. Errors spread their effects using the same channels that the organization uses in its normal operations. Hence, if a serious design flaw appears, it could affect the whole fleet of a manufacturer – that is, every single plane whatever its size or range.

Furthermore, several events, one of them severe enough for a plane to be written off, have occurred because fatigued or hurried pilots confused the plane they were flying and set the wrong take-off parameters. Although the computer will not accept the input of non-valid parameters, a heavy plane fueled for a short flight can have a similar weight to a lighter plane fueled for a long flight.

Additionally, a change in the air traffic control (ATC) plans, allowing a take-off from a shorter runway or from an intersection in a long one[3] can cause the pilot to miss the introduction of the new power values. Therefore, despite the existence of systems such as the Runway Awareness and Advisory System (RAAS), the avoidance of this type of error is not always possible.

In this context, the close, interlinked, efficiency-driven relationships among manufacturers, operators and regulators appear.

The role of regulators is not to spoil an industry through close control but to introduce discipline without suffocating the regulated industry. In a wider sense, they are the representatives of citizens in a field where those citizens lack basic knowledge about a widely used product, whose use can even imply serious life risk. In other words, regulators are supposed to bring into the system the "superior rationality of the experts", as Beck suggested.

Operators compete among them, always asking for changes aimed at efficiency improvement. That, as shown, forces manufacturers to offer remarkably similar products. If manufacturers can supply cheaper products in every dimension – that is, when bought, when operated and when maintained – and, at the same time, they are similar enough to reduce stocks for maintenance and to decrease training time between models or make possible the use of the same crews for different planes, that would be a perfect world.

This perfect world is almost here, but it introduces a new risk coming from the tightly coupled organizational model that it implies. How to deal with a situation where something unusual and affecting a considerable number of airplanes occurs?

The experience of grounding the DC10, the B787 or, more recently, the latest B737 due to design flaws is already known, but what if a major design failure affects every single plane coming from one of the big manufacturers? Similarity or commonality could make that scenario possible.

The Learning Process: How the System Evolves

DC10 design is a milestone for learning in Aviation: A bad design in the cargo door, with potential for a major accident – something known even before the plane started to fly passengers – was considered after the first serious event as something that *could be* improved, not as a mandatory airworthiness directive – that is, something to fix before the plane was allowed to fly again.

The difference between both options is truly relevant even from the economic side: If the plane must change to be airworthy, the related costs are on the manufacturer. A suggested improvement, on the other side, is on the operator. The relevance of this difference would become clear in the DC10 case.

> A cargo door suddenly opened, causing explosive depressurization, but the pilot miraculously managed to land the plane (AA96). The suggested change involved some reinforcements and an informative plate, but the basic design remained.

Two years later, the same problem reappeared on a Turkish Airlines flight, but this time the occupants of the plane were not so lucky: The depressurization broke the control wires and the plane crashed with 325 people on board (TK981).

The investigation revealed to what extent the problem was known during the design. A dismissed written report, including tests clearly showing the defect and its potential effects, became public.

After the first event, something known as a "gentlemen's agreement" (as the official report called it) between manufacturer and regulator had downgraded the importance of the design problem. After a major event like TK981, these facts could not remain hidden and many passengers refused to fly the DC10.

In some ways, this event sealed the fate of its manufacturer, McDonnell Douglas. They accelerated the design of a substitute, MD11, that was born with many defects coming from that rushed design that would lead to new events. Additionally, the new MD11 was hard to distinguish from the old DC10 to untrained eyes.

DC10 had still more cases related to design: The accident of American Airlines 191, still the worst Aviation event in U.S. history if we except terrorist attacks, mixes design features and poor maintenance practices.

During take-off, one of the engines detached from the plane because of a poor maintenance practice that weakened the link between engine and wing. Let's say that everything was still "normal" in the sense that, in theory, a fatal risk should not have come from that separation.[4]

However, DC10 had a new feature: It had suppressed the mechanical blockage of the flaps and slats. Flaps and slats remained in the extended position through hydraulic means, and, to get approval, the manufacturer had to show that they remained in position even if an engine stopped. The manufacturer showed that and the plane became certified.

Facts would show that an engine stop was not the only, nor the worst, possible scenario for that system, and a different and worse option remained untested: the consequences of the separation of an engine during flight. The separation of an engine cuts all the wires and pipes – bringing electricity, hydraulic fluid, fuel or whatever – linking the engine with the wing, leading to loss of fluids.

That is exactly what happened. The plane lost hydraulic fluid, and the flaps and slats, kept in position by that fluid, retracted at the most critical moment – that is, during take-off.

Additionally, a warning system, the stick-shaker, was linked to the engine that disappeared and, hence, it became inactive after the engine detached from the plane. All the information available to the pilots, who could not see the engines from the cockpit, was coherent with an engine stop. Then, they reacted as trained, increasing the angle to gain altitude. By doing so, since the plane had lost lift due to the retraction of the flaps and slats, they provoked a stall and the plane crashed close to the airport.

Tests performed in a simulator after the accident would show that, had the pilot known the nature of the problem, the plane would have been recoverable. However, with the available information, it was not.

After this accident, one might think that the use of hydraulic fluid to keep flaps and slats in the extended position would not be a sound idea. However, this feature remained in DC10 by adding a valve to keep the system working if a hydraulic loss occurred. The successor of DC10, MD11, appeared with an *improved* design, with the standard mechanical blockage that the DC10 had given up.

Another weak answer from the regulator came with Northwest 255.

The plane was a different model (MD82) from the same manufacturer. In this case, the configuration alert failed, and the plane tried to take off with the flaps retracted without any warning in the cockpit.

The "human error" stamp in the conclusions only required a change in procedure – sent only to the present operators of the plane – and it was issued instead of an airworthiness directive requiring a design change.

Twenty-one years later, the Spanair JKK5022 accident occurred for exactly the same reasons, making it evident that the Northwest lesson had been ignored. Spanair received no information about the procedure since they were not flying this model when Northwest 255 took place.

There are no elements to establish the existence of another "gentlemen's agreement" between FAA and McDonnell Douglas, and, if so, it should be difficult to prove it after such a long time, but the behavior of the manufacturer regarding the DC10 and MD82 design faults was very similar.

In both cases, Northwest and Spanair, all the fingers pointed to human error. However, after the Spanair case, a search in the ASRS database showed more than 400 reports where pilots forgot to set the flaps. The number was high enough and the causes leading to it were known well enough to make sure that the configuration alert worked properly.[5] It didn't in 1987 and, again, it didn't in 2008.

Supposedly, learning from accidents does not mean trying to avoid the last one, since the causality chains are so complex that a full repetition is extremely improbable. However, that means that learning should not address *only* the eventual repetition of the accident but also the root causes leading to different accidents, including the ones that have already occurred. In the unlikely event of a recurrence of a former accident, we have strong reasons to conclude that something has been lost in the process.

Finally, another important case (AF447) that could be related to design happened in 2009.

A sensor froze over the Atlantic and it started to send incorrect information to the systems of the plane. The pilots became confused in the resulting situation and the plane fell 10,000 m to crash in the ocean while both engines were working as well as the flight controls.

Some elements added confusion to the effects of the frozen sensor: a silent stall warning that, against intuition, did not mean a situation improvement, but just the opposite. When the plane speed was out of the valid interval for this parameter, the system would remain silent. Meanwhile, it would go off again when the plane started to recover but still at very low speed values. Then, an absurd situation appeared, where warning would mean recovery while silence would mean danger.

Additionally, the sidestick did not give any kinesthetic feedback, informing the pilot through pressure that something was wrong. Furthermore, the sidesticks of both sides were independent. Hence, in a dark cockpit, the pilot not flying cannot see the other pulling the sidestick, making recovery impossible. Additionally, the noise and the stress can easily lead to one losing the indication of input discrepancy, when every pilot is actioning his flight stick in a different direction. The accident was officially attributed to lack of training, our old friend, despite the complaints of the operator Air France.

Some experts, such as Sullenberger, the pilot in the US1549 case, said that this accident would be less likely with a Boeing model[6]: The fly-by-wire system – used in all Airbus fleets except the old A300s and A310s – does not have a mechanical link between flight controls and flight surfaces, as mechanical or hydraulic systems have. It would not be expected, then, to have any kind of feedback through force transmitted to the hand.

When Boeing introduced fly-by-wire in the B777 and B787 models, hydraulic feedback was kept with the explicit objective of providing feedback through pressure, similar to that in a Boeing 767, and, of course, to have a hydraulic backup in the event of a total failure of the fly-by-wire system. Then, in the case of Boeing, this is an added feature instead of a side-effect of the old system, as designed in planes previous to the fly-by-wire generation.

That means greater mechanical complexity in the Boeing case, while Airbus simply accepted the situation; that is, since there is no mechanical link between flight controls and controlled surfaces, nobody should expect a kinesthetic feedback in the flight controls from these surfaces.

The difference, then, between these manufacturers is that Boeing dealt with the pressure on flight controls as a highly appreciated information resource or, if preferred, a tactile alert. The pressure felt through the flight control can tell the pilot how the plane is flying, but sidesticks in Airbus models (similar to joysticks used for video games) don't provide pressure and, therefore, don't provide the information coming from that pressure.

When an accident happens, using hindsight, it is easy to say what the pilot *should* have known. Sometimes the pilot is supposed to know on-the-go things as counterintuitive as a silent stall warning being worse than a noisy one. If the pilot did not know what was going on, we call it "lack of training" instead of poor design, especially if it can affect many planes.

Lack of training, a label already criticized here, is very often a circular explanation that helps to keep the situation as it was before a major event:

Someone close to the event performs a task inadequately or does not perform a required task because of missing knowledge or training. Then, the event is attributed to this lack of training or knowledge. A few more lines in the training manual will solve the problem and the event is not supposed to appear again.

Of course, this extended practice does not answer the main question, which we will try to answer in the following chapters:

Why was someone without the required training performing the task that made this lack of training clear? The label "lack of training" suggests that the training managers were lazy, incompetent or both, but very often that is not the cause.

Simply put, nobody foresaw the event and, hence, nobody was trained for it. Then, some sophisms must be performed before calling it "lack of training", the most common of them being trying to deduce it from broad categories; that is, if a regulation says that the pilot must always keep control of the plane, barring physical impossibility, and the pilot loses control, the answer is lack of training.

There is not a homogeneous and internally consistent training syllabus pointing to which things must be known or trained in advance to manage every imagined event. Instead, the syllabus changes through new and unforeseen events, adding contents to the required training, but, at the same time, and based in generic training requirements, the unforeseen event is attributed to lack of training.

Very often, had the pilot known something, the event would not have taken place. However, pointing to lack of training and changing the syllabus to include the newly discovered event is an efficient – but not always effective – way to deal with the next similar event. At the same time, it is entirely useless for preventing unforeseen events.

Take as an example the official report of an extremely complex event such as XL-888T, where two out of three angle-of-attack sensors became frozen almost at the same position, leading to bizarre behavior by the automatic systems. Although the flight is recognized in the official report not to be a test flight and the crew had the adequate licenses and qualifications, they were supposed in that same report not to have the required skills to perform that kind of flight.

What should be the technical skills, experience and methods to perform a flight recognized as "not a test flight"? Does it mean that the accident would not happen with a different crew?

One of the most repeated sentences in EASA CS25 (2018) is this:

> No ... to determine the data required by this paragraph may require exceptional piloting skill or alertness.

This insistence in different sections of EASA CS25 is a reminder for the designers that systems – and the whole plane – must be suitable for common people with the right training. If, as made explicitly clear, the report does not call it test flight and the crew had the required licenses and qualifications, why the remark pointing to lack of training?

However, this situation is far from being uncommon. The system always looks for the most efficient way to prevent the recurrence of an event. Although cases like this should leave a question mark over the design, lack of training or human error are usually the preferred labels to close an open issue.

Major design changes are, as far as possible, avoided, at least for planes already certified and flying. Regulators try to introduce the learning obtained through different events. As an example, a major accident happened in 1992 in Mount St. Odile, France.

> The pilots committed a mistake in the setting of a flight instrument that could be set alternatively to an angle of descent or to a rate of descent. The interface design made the confusion easy and, of course, setting an angle of descent of 4 degrees is vastly different from setting a rate of descent of 4,000 ft./min. So the plane was descending at a widely steeper angle than intended and crashed during approach against Mount St. Odile.

The regulation that applies to the design of the instrument that caused the error is CS25.1329, "Flight Guidance System". Going back in time, from the version updated in March 2018 to the old JAR25.1329 issued in 1989 and then called "Automatic Pilot System", something interesting can be found.

Despite the change in the heading, there are no many major changes in the main body of the rule, even though one of the changes points directly to the root of this accident (EASA CS25, 2018).

> (i) The flight guidance system functions, controls, indications, and alerts must be designed to minimise flight crew errors and confusion concerning the behaviour and operation of the flight guidance system. Means must be provided to indicate the current mode of operation, including any armed modes, transitions, and reversions.

The expression "flight crew errors" simply did not exist in any paragraph of the older version and another subtler change can be found.

When something is noticeably clear for the regulator, the regulation is clear and prescriptive. When something is fuzzy, it is not uncommon to find negative sentences such as "Each pilot compartment and its equipment must allow the minimum flight crew (established under CS 25.1523) to perform their duties *without unreasonable concentration or fatigue*" or sentences whose meaning is clearly negative, such as *"minimize flight crew errors"*.

A prescriptive approach gives the affected parts clear guidance about what to do and when a design is compliant with the rule or not. An approach based on an abstract goal – number of errors, fatigue or any other factor – is always open to the interpretation of a future event by forecasting the past – that is, what the involved people should know.

Afterward, this could show a lack of compliance leading to an airworthiness directive. In other words, the same dynamics of rules found with "lack of training" appears in other fields through fuzzy or goal-oriented regulations.

Then, regulators try to learn from new events while protecting the regulations from issues that can be fuzzy. Of course, the fuzziest of them are usually those related to human factors.

A last example to clarify this point: Regulators have made important efforts to clarify the minimum amount of fuel that a plane must transport. The subject is critical and clear enough to require further explanation. Let us suppose that a regulator issues a new rule where the required amount of fuel should be "the necessary". This choice of words would be perfectly acceptable from the regulator's point of view, since any future event related to fuel exhaustion would mean that the operator did not follow the rule; that is, the accident itself would become the proof of noncompliance. However, a rule like that would open the door to organizational pressures due to the lack of guidance and a clear prescription.

There is another major source for regulations: the evolution of technology. Taking as an example the same rule, CS25.1329, we can find that differences are much bigger in the Acceptable Means of Compliance (AMCs) than in the main body of the rule.

While the AMC equivalent for JAR25.1329 had seven pages, the latest version of AMC25.1329 has 72 pages. However, while the paragraph in the main body of the rule is easy to trace to a major event, the AMC comes from major technological changes that occurred in the time between both versions of the regulation. The newer version tries to define requirements for systems that did not exist when the older one was issued, or, if so, they were in an early phase.

Summarizing, Aviation has grown over the years both in the volume of operations and safety levels. The main manufacturers have been trying to improve efficiency while claiming to maintain or improve safety levels. Any radical change that is able to increase efficiency is often rejected by those who do not introduce it. However, since manufacturers are pressed by operators to be increasingly efficient, these changes, time and again, are finally imitated, accepted and adopted.

This process creates a situation where new models coming from the main manufacturers are increasingly alike, but since development times are longer and costs are higher, they try to keep their old models alive for a long time. One of the ways to keep an old design alive is to dismiss any change that could render the models already on the market obsolete.

At the same time, an important flow of information is coming from events of varying relevance. This information can lead to change in regulations and, therefore, in technology and systems. However, when a serious event affects a plane in a way that could lead to major changes, there is a trend to minimize, pointing to training, human error or minor changes.

This trend decreases when the problem affects newly designed planes. In this case, changes are easier to perform since they don't affect many units already flying. Changes in regulations don't appear only in the guise of new and clear prescriptions. Very often, negative statements based on abstract goals are used. Then, nobody is fully sure of being compliant and it brings, as a side-effect, the practice of keeping a certified plane flying with minor changes.

New models are delayed by increasing development time and costs,[7] leading to a paradox: Regulations can be seen as a repository for organizational learning and, at the same time, this repository can delay the advancement instead of helping it.

Enlightened Despotism in Aviation

Operators, regulators and manufacturers are the main organizational pieces of the entire system. They and the relationships between them can give us a good idea of how the whole system works. These relationships are desirable but not always functional.

Operators want planes to be as efficient as possible in every single dimension. They press manufacturers in that direction, but at the same time manufacturers know that adventures in design can be very costly. Regulators try to keep Aviation safe without making new developments aimed at profitability impossible.

Additionally, they must manage new challenges, such as increasing air traffic, demanding new ways to control a crowded airspace, and, even more so, they have their own internal challenges such as harmonizing rules coming from the main regulators, EASA and FAA, to avoid having the rules become a confusing forest.

The common use of generic or non-prescriptive statements drives designers to use standards from various sources – Engineering, the Military and others – to find specific guidance instead of generic goals or things to avoid.

The problem with this practice, beyond the confusion linked to having standards taken from various sources, is that these standards apply to specific environments. For example, Military standards must consider the operation under restrictions such as partial physical destruction, exceptional work conditions or lack of resources. Receiving guidance from these fields often leads to overcompliance or to designs that are unsuitable for the Commercial Aviation environment.

As an example, this text from MIL-STD-1472G (US Department of Defense, 2012) should not apply to Commercial Aviation.

> Shutters having closure and reopening times appropriate for each application may be provided in lieu of fixed filters to protect the observer exposed to flashes from weapon systems, lasers, or other bright light sources.

Perhaps the frequency of incidents coming from laser pointers used near to airports could lead to the adoption of this military-oriented design rule, but, in any case, it was intended as a protection against expected activities in an hostile zone.

Beyond these organizational dynamics related to rules and where they come from, individual events – or black swans – can make new demands appear. For instance, the AF447 or MH370 cases revealed to a surprised public that a plane can disappear. Then, common people suddenly discovered that a flying plane is not always on the screen of a controller. This common piece of knowledge for specialists became an unwelcome surprise for lay people.

MH370 brought that surprise to a different level, even for many people familiar with Aviation. Immediately, something supposedly guaranteed long ago became an acceptability requirement: to know always and in real time where a plane is.

This is a closed system where, from time to time, details leak to passengers and the public, while the main actors decide on behalf of them, comforting users with, "Trust me, this is safe."

From a technical and managerial point of view, this behavior is adequate, especially in a field where common users cannot even explain how planes fly. However, from the social and organizational points of view, it is risky.

Previously unknown decisions – good or bad, clear or suspicious – can lead to rejection if the media or the internet present a major event – even unfairly – as being driven by a hidden decision or a "gentlemen's agreement". Information sharing should be practiced, rather than simply keeping discussions among colleagues and limiting them to a safe place.

Issues such as ETOPS certification and its implications should be more openly discussed. Deciding on subjects such as that among manufacturers, regulators and operators will drive usually to a technically correct decision. However, while keeping the lay consumer out of the decision can be preferable in the short term, it would be naïve to dismiss the potential social impact of a major accident.

The quality of technical decisions can be right in terms of the assumed risk. Even so, reservations could remain about some tricky regulations, apparently aimed at saving the regulation itself by creating a fuzzy environment, instead of helping the users.

However, as previously commented on, people might reject even good technical decisions if a major event, seen through the lenses of the lay public and journalists, appears to be a major scandal or an undesired "trending topic". Main actors should inform people that flying implies risk, albeit minimal, instead of keeping them ignorant about this single fact.

These main actors could argue – and they would be right – that common users are not qualified to evaluate technical decisions. However, the same people arguing this point would be outraged by the behavior of a physician making decisions about their health without giving them a say.

The present system, then, gives comfort to those making decisions in a closed environment, but this comfort could have a high, albeit deferred, cost.

The DC10 case illustrates this point: The so-called "Applegate memorandum", written before Turkish Airlines 981, showed that "it seems to be inevitable that, in the twenty years ahead of us, DC10 cargo doors will come open and I would expect this to usually result in the loss of the airplane". This warning was disregarded at an extremely high cost: McDonnell Douglas never recovered from DC10 and one of the biggest names in Civil and Military Aviation disappeared from the market.

Other decisions, even if they not so clearly wrong as those related to DC10, are subject to a similar threat. If a major event related to a technical decision unknown to the public occurs, will the organizational network, as it works today, stop the tsunami that could result from it?

Over the last decade, the two aforementioned lost planes (AF447 and MH370) made it clear that we do not always know where a plane is. That triggered changes in something widely known inside Aviation but unknown to a public that never suspected it to be possible.

Could we imagine, for instance, the public impact of a full loss of power over an ocean, leading to a major event? That has already been seen with events caused by fuel exhaustion or volcanic ashes; it is not due to a risk assessment about the reliability of the engines. What about the impact of lithium batteries on planes where electricity plays a key role? A major cabin smoke event? A massive electrical failure? A software failure?

Every flight has a demonstration about emergency exits and oxygen masks, but does anyone inform passengers wearing short trousers and sleeves of the impact of depressurization on the temperature if they happen to be far from an airport? The list of potential events fully unknown to the passengers would be much longer.

People are kept ignorant about many basic facts related to their own safety. This practice is preferable to Aviation managers, and understood by Aviation as a whole, but it comes with a major, albeit avoidable, risk: negative public perception if people feel misinformed or betrayed about known dangers and decisions made in the dark.

We have focused on the design of planes and the side-effects related to this factor. However, we cannot forget that ATC suffers a similar dynamic, due to an increasingly crowded airspace.

Both aspects are closely related, since many of the resources to manage a crowded airspace imply that pilots cannot fly planes manually in that space due to the precision requirements.

These facts shape a scenario where some technical decisions, present and expected in the short term, should be seen through the eyes of the public before implementing them. Some others should be informed in clear terms to hold back disinformation and sensationalism if a major event happens.

By not doing so, a major issue appears: an "enlightened despotism" that makes life easier but is forced to decree some events as black swans, even if they are foreseeable.

Notes

1. If an engine stop happens after V1 (decision speed), the plane must be able to take off with the remaining power. That means 75% of nominal power in a four-engine plane and 50% of nominal power in a twin plane. The available power decreases from the nominal level because of the thrust asymmetry that a failed engine implies. This effect is still bigger in twin planes than in four-engine planes.
2. https://services.airbus.com/training/flight-crew/02-pilot/cross-crew-qualification-reduced-type-rating-course
3. https://dohanews.co/report-qatar-airways-fires-pilots-involved-in-miami-takeoff-incident
4. Some years later, El Al 1862 took place in Amsterdam. The plane, a B747, lost an engine. It destroyed a good part of the leading edge, hit the other engine in the same wing, causing it to separate too, and the plane crashed.
5. Actually, the configuration alert is in the minimum equipment list on large airplanes.
6. https://youtu.be/kERSSRJant0
7. The introduction of a new model and the certification process is so costly that companies such as the Chinese COMAC, instead of running the full process, decided to certify their first plane to fly only inside China.

Bibliography

Ahlstrom, V., & Longo, K. (2003). Human Factors Design Standard (HF-STD-001). Atlantic City International Airport, NJ: Federal Aviation Administration William J. Hughes Technical Center.

Air Force, U.S. (2008). *Air Force Human Systems Integration Handbook*. Air Force 711 Human Performance Wing, Directorate of Human Performance Integration, Human Performance Optimization Division.

Australian Transport Safety Bureau (2011). Investigation Number: AO-2009-012 Tailstrike and runway overrun: Airbus A340-541, A6-ERG, Melbourne Airport, Victoria, 20 March 2009.

Aviation Safety Network (2007). Turkish Airlines Flight 981. Accident report https://reports.aviation-safety.net/1974/19740303-1_DC10_TC-JAV.pdf [in French].

Aviation Safety Reporting System (ASRS) (2019). ASRS Database Online. https://asrs.arc.nasa.gov/search/database.html.

Bainbridge, L. (1983). Ironies of automation. In: *Analysis, Design and Evaluation of Man–Machine Systems 1982* (pp. 129–135). Pergamon, Oxford.

Beaty, D. (1991). *The Naked Pilot: The Human Factor in Aircraft Accidents.* Airlife.Shrewsbury, England

Beck, U. (1992). *Risk society: Towards a new modernity* (Vol. 17). Sage.

Beck, U., & Rey, J. A. (2002). *La sociedad del riesgo global.* Madrid, Spain: Siglo Veintiuno.

Bennett, K. B., & Flach, J. M. (2011). *Display and Interface Design: Subtle Science, Exact Art.* Boca Raton, FL: CRC Press.

Bureau d'Enquêtes et d'Analyses pour la sécurité de l'aviation civile (1993). Official report into the accident on 20 January 1992 near Mont Sainte-Odile (Bas-Rhin) of the Airbus A320 registered F-GGED operated by Air Inter.

Bureau d'Enquêtes et d'Analyses pour la sécurité de l'aviation civile (2010). Report on the accident on 27 November 2008 off the coast of Canet-Plage (66) to the Airbus A320-232 registered D-AXLA operated by XL Airways German.

Bureau d'Enquêtes et d'Analyses pour la sécurité de l'aviation civile (2012). Final report on the accident on 1st June 2009 to the Airbus A330-203 registered F-GZCP operated by Air France flight AF 447 Rio de Janeiro–Paris.

Business Insider (April 2, 2018). If you have a Tesla and use autopilot, please keep your hands on the steering wheel. https://www.businessinsider.com/tesla-autopilot-drivers-keep-hands-on-steering-wheel-2018-4?IR=T.

Comisión de Investigación de Accidentes e Incidentes de Aviación Civil (2011). A-032/2008 Accidente ocurrido a la aeronave McDonnell Douglas DC-9-82 (MD-82), matrícula EC-HFP, operada por la compañía Spanair, en el aeropuerto de Barajas el 20 de agosto de 2008.

Commercial Aviation Safety Team (CAST) (2008). Government working group report. Energy state management aspects of flight deck automation: Final report.

DoD, U.S. (2012). Department of Defense Design Criteria Standard: Human Engineering (MIL-STD-1472G). Washington, DC: Department of Defense.

Dörner, D., & Kimber, R. (1996). *The Logic of Failure: Recognizing and Avoiding Error in Complex Situations* (Vol. 1). New York: Basic Books.

EASA CS25 (2018). Certification specifications for large aeroplanes: Amendment 22. https://www.easa.europa.eu/certification-specifications/cs-25-large-aeroplanes.

Fielder, J. & Birsch, D. (1992). *The DC-10 Case: A Study in Applied Ethics, Technology, and Society.* New York: State University of New York Press.

Fischhoff, B., Lichtenstein, S., Slovic, P., Derby, S. L., & Keeney, R. (1983). *Acceptable Risk.* Cambridge, UK: Cambridge University Press.

Fleming, E., & Pritchett, A. (2016). SRK as a framework for the development of training for effective interaction with multi-level automation. *Cognition, Technology & Work,* 18(3), 511.

Hale, A. R., Wilpert, B., & Freitag, M. (Eds.) (1997). *After the Event: From Accident to Organisational Learning.* New York: Elsevier.

Harris, D. (2011). Rule fragmentation in the airworthiness regulations. *International Conference on Engineering Psychology and Cognitive Ergonomics EPCE 2011: Engineering Psychology and Cognitive Ergonomics pp. 546-555.*
Hubbard, D. W. (2009). *The Failure of Risk Management: Why It's Broken and How to Fix It*. John Wiley.
Komite nasional keselamatan transportasi Republic of Indonesia (2018). Preliminary KNKT.18.10.35.04 aircraft accident investigation report PT. Lion Mentari Airlines Boeing 737-8 (MAX); PK-LQP Tanjung Karawang, West Java, Republic of Indonesia, 29 October 2018.
Leveson, N. (2004). A new accident model for engineering safer systems. *Safety science*, 42(4), 237–270.
Leveson, N. (2011). *Engineering a Safer World: Systems Thinking Applied to Safety*. Cambridge, MA: MIT Press.
Leveson, N. G. (1995). *Safeware: System Safety and Computers* (Vol. 680). Reading, MA: Addison-Wesley.
Leveson, N. G. (2004). Role of software in spacecraft accidents. *Journal of Spacecraft & Rockets*, 41(4), 564.
Maurino, D. E., Reason, J., Johnston, N., & Lee, R. B. (1995). *Beyond Aviation Human Factors: Safety in High Technology Systems*. Aldershot: Ashgate.
National Research Council (1997). *Aviation Safety and Pilot Control: Understanding and Preventing Unfavorable Pilot-Vehicle Interactions*. Washington D.C.: National Universities Press.
National Safety Council (NSC) (2017). NSC motor vehicle fatality estimates. Statistics Department, NSC. https://www.nsc.org/Portals/0/Documents/NewsDocuments/2018/December_2017.pdf.
National Transportation Safety Board (1973). NTSB/AAR-73/2 American Airlines McDonnell Douglas DC10-10, N103AA near Windsor, Ontario, Canada, June 12, 1972.
National Transportation Safety Board (1979). NTSB/AAR-79/17 American Airlines DC10-10, N110AA Chicago, O'Hare International Airport, Chicago, Illinois, May 25, 1979.
National Transportation Safety Board (1988). NTSB/AAR-88/05 Northwest Airlines, Inc. McDonnell Douglas DC-9-82, N312RC Detroit Metropolitan Wayne County Airport Romulus, Michigan, August 16, 1987.
National Transportation Safety Board (1990). NTSB/AAR-90/06 United Airlines Flight 232 McDonnell Douglas DC-I0-10 Sioux Gateway Airport Sioux City, Iowa, July 19, 1989.
National Transportation Safety Board (2010). NTSB/AAR-10/03 Loss of thrust in both engines after encountering a flock of birds and subsequent ditching on the Hudson River, US Airways Flight 1549 Airbus A320-214, N106US Weehawken, New Jersey, January 15, 2009.
Netherlands Aviation Safety Board (1992). Aircraft accident report 92-1 1 El Al Flight 1862 Boeing 747-258F 4X-AXG Bijlmermeer, Amsterdam, October 4, 1992.
Perrow, C. (1972). Complex organizations: A critical essay. New York: McGraw-Hill.
QCAA Qatar Civil Aviation Authority (2015). Preliminary accident report A7-BAC. https://www.caa.gov.qa/sites/default/files/Preliminary%20report%20QR778%20Miami_v3.pdf.
Reason, J. (1997). *Managing the Risks of Organizational Accidents*. Aldershot: Ashgate.
Reason, J. (2017). *The Human Contribution: Unsafe Acts, Accidents and Heroic Recoveries*. Boca Raton, FL: CRC Press.

Rudolph, J., Hatakenaka, S., & Carroll, J. S. (2002). Organizational learning from experience in high-hazard industries: Problem investigation as off-line reflective practice. MIT Sloan School of Management working paper 4359-02.

Sagan, S. D. (1995). *The Limits of Safety: Organizations, Accidents, and Nuclear Weapons*. Princeton, NJ: Princeton University Press.

Sánchez-Alarcos Ballesteros, J. (2007). *Improving Air Safety through Organizational Learning: Consequences of a Technology-Led Model*. Aldershot: Ashgate.

Taleb, N. N. (2007). *The Black Swan: The Impact of the Highly Improbable* (Vol. 2). New York: Random House.

Tesla (2018). Full self-driving hardware on all cars. https://www.tesla.com/autopilot.

The Malaysian ICAO Annex 13 Safety Investigation Team for MH370 (2018). Safety investigation report, Malaysia Airlines Boeing B777-200ER (9M-MRO), 08 March 2014.

Transportation Safety Board of Canada (1998). Report number A98H0003: In-flight fire leading to collision with water, Swissair Transport Limited, McDonnell Douglas MD-11 HB-IWF Peggy's Cove, Nova Scotia, 5 nm SW, 2 September 1998.

US Department of Defense. (2012). MIL-STD-1472G. Design Criteria Standard: Human Engineering.

Walters, J. M., Sumwalt, R. L., & Walters, J. (2000). *Aircraft Accident Analysis: Final Reports*. New York: McGraw-Hill.

Wells, A. T. (2001). *Commercial Aviation Safety*. New York: McGraw-Hill.

Winograd, T., Flores, F., & Flores, F. F. (1986). *Understanding Computers and Cognition: A New Foundation for Design*. New York: Addison-Wesley.

Womack, J. P., Womack, J. P., Jones, D. T., & Roos, D. (1990). *Machine That Changed the World*. New York: Simon and Schuster.

2

Event Analysis as an Improvement Tool

As shown, air safety has exhibited an impressive performance in its organizational learning process, if measured by improvement ratios. The improvement rate started to decrease around 1975, and the resulting curve was mainly asymptotic.

In developed countries, the accident rate is so low that a single major accident will show a peak in the curve corresponding to the year when that accident happened.

We could say that Aviation – defined as an activity whose main players are manufacturers, operators and regulators – has improved through a learning process, but, at the same time, the learning process itself has changed over time.

Originally, the learning process was event driven. In early Aviation, learning was aimed at avoiding the repetition of events with potentially or factually negative outcomes.

The events can be understood as deviations from the expected action. Such deviations show shortcomings that, once identified, should lead to the implementation of the improvements necessary to avoid their repetition. From the beginning of Commercial Aviation, identifying the causes of an event has been a basic improvement tool.

Although event analysis is still relevant, there are two related factors that have encouraged the involved organizations to include different inputs.

- First, the size of the planes has been increasing over time. A single accident can be a major disaster, affecting up to 600 people, if the accident involves a single plane. Therefore, waiting for major events to occur in order to learn is not acceptable.
- Second, the number of major events has been decreasing and, hence, the number of opportunities to learn from them has decreased too.

Then, there are two options, both widely used:

1. Anticipate potential events before they happen. This learning is included in the design phase, whether it is a design of the plane, a procedure or any task coming from it.
2. Process information about minor events through reporting systems.

Therefore, an event does not need to be major to deserve analysis. Major events are prevented, as much as possible, in the design phase while minor events are managed as precursors. Hence, minor events are analyzed in terms of the likelihood that they will lead to a major one.

The role of event analysis in safety improvement can be analyzed from a dual perspective:

1. Potential and uses of event analysis.
2. The event-based learning cycle.

Potential and Uses of Event Analysis

The potential severity of an event has grown at the same rate as the size of airplanes. Because of this, its investigation occupies a key place in the whole safety system. A single major event can be the origin of important technological or procedural changes.

Significant data about this path of learning are shown by cases such as the only accident of a Concorde.

> The accident happened after a piece of metal, detached from a plane that took off before Concorde, produced a tyre blow-up almost at take-off speed. Rubber pieces were launched toward the lower side of the wing and to the air intake of the engines. That resulted in the puncture of a fuel tank and the stopping of two engines during take-off.

This accident first resulted in the replacement of fuel tanks and, afterward, it provided the perfect justification to retire a plane that was very expensive to maintain and operate while, at the same time, it was the only supersonic commercial plane flying.

In the same fashion, repeated cases of engines being mistakenly switched off during engine fires would lead to the requirement to confirm which engine had problems before applying extinguishers; cases such as that of the aforementioned St. Odile would lead to relevant changes in the regulations regarding instrument design.

The role of event analysis as a learning tool has wide recognition, coming from fields as varied as the strictly technical, the philosophical or even the psychological.

Maturana and Varela (1987) point out that the failures of machines constructed by man are more revealing about their effective operation than the descriptions we make of them when they do not fail. From a different point of view, Maturana and Varela's statement is a description of the clinical method, where conclusions are obtained about normal functionality by using the abnormal as the starting point.

Supposedly, this method is more oriented to practices other than engineering: Planes are manufactured by people that are supposed to have perfect knowledge about the object of their work. It should be a very different situation with naturally unpredictable elements. However, that is not the whole truth: Designers are still surprised both by effects related to users' interactions and by unforeseen or miscalculated problems of the design itself. Interactions must be designed as carefully as any other element, and error proneness can be introduced through poor design.

About the design itself, cases such as that of Comet planes, which will be commented on later, show that important variables can be missed in the design process. The analysis of events can bring these elements to the surface.

In the philosophical field, Karl Popper contributed with his falsifiability principle. Popper insisted that the real proof for any hypothesis should not be its feasibility; the real proof should be showing that it could not be false. In this way, Popper tried to attack the common trend of accepting at face value the most plausible option.

In practical terms, the use of this principle requires one to devote the same effort to the dismissed options as to the preferred one. This is one way to avoid "tunnel vision", which can bring certainty but not necessarily the right guess.

Therefore, event analysis is especially useful for highlighting the unsound parts of a process, if there is information enough for its reconstruction. In the case of Commercial Aviation, the level of destruction in some major accidents makes gathering information difficult. Therefore, specific resources will be required for that.

Sometimes, the investigation of an event has led to the production of high-quality experimental designs, high costs and even personal risk in the search for the cause of an accident. Possibly the most representative case may be the aforementioned De Havilland Comet. Comets – the first commercial jets in the world – were exploding at cruising altitude without anyone having a single clue about why it was happening.

> In the first phase of the investigation, flights were carried out at the altitude at which the explosions had occurred, taking technicians with parachutes and oxygen masks as passengers. However, on those flights the airplane was not pressurized and nothing abnormal could be observed since, as it would be shown later, pressurization was the root cause of the accidents. Once it was suspected that the failure was due to the pressure exerted on the fuselage by the pressurization system, investigators kept the airplane on the ground, filling the fuselage with water under pressure and then emptying it repeatedly.
>
> One simple explanation of the phenomenon is as follows: Let's suppose that we repeatedly inflate and deflate a balloon inside a bottle. If the material or the thickness of the bottle are not sufficient, a point will be reached when the pressure of the balloon will break the bottle; the Comet had large windows that permitted passengers better views than on other airplanes, but those windows reduced the strength of the structure until it broke.

Most probably, an experiment like the one detailed would not be required today, as a computer could run the simulation instead, but with the available resources at that moment, they found that the action produced in the experiment the same structural failure that took place on the planes, due to metal fatigue.

Changes were made, but the newly acquired knowledge went far beyond the Comet. It was applied to new jets, whoever was manufacturing them, that without this knowledge would have been subject to the same fate.

Other risky and relevant experiments have been performed on the effects of wake turbulence – caused by the movement of a large airplane – or of icing on ATR planes.

In the first case, tests were carried out during flight with two airplanes of different size, where the smaller – a Boeing 737 – was deliberately introduced into the wake left by a larger plane to check the effects on its controllability. The experiment permitted the elimination of this possibility in an investigation. Years later, in one of the longest investigations in Aviation, a failure would be found in the design of the rudder of the airplane.

Very similar to this experiment was the one carried out to determine the effect of ice on ATR airplanes. One airplane in flight released frozen water onto the ATR to check the effects produced on it and whether this could explain the sudden loss of control in the accidents. In this case, the experiment led to the conclusion that the accumulation of ice on the ailerons resulted in a sudden and difficult-to-control deflection.

Smaller experiments are regularly performed in accident investigations where the parameters of the flight, as recorded, are reproduced in a simulator. Even though this is not always a valid solution, it helps to clarify many accidents.

The importance of the reconstruction of an event led to the construction of devices that would allow the information to survive an event with a high level of destruction. Over time, this field has evolved to a point that few accidents have remained unexplained in Commercial Aviation for some time.

Obviously, learning from events requires these events to happen. A major event makes people uneasy until explained. Then, there is a strong motivation to know what happened. Minor events are subject to very different rules:

Unless the event is a very close call, the motivation to report and investigate decreases. This effect increases if the event is in any way the responsibility of the reporter or the reporter is afraid of any kind of revenge.

This fact is important since it is commonly accepted that events, whether or not they lead to a major accident, can share a root cause. It's generally accepted that the ratio of minor events to accidents is 300:1; that is, for approximately every 300 minor events there is an accident.

Under this rationale, capturing minor events should prevent major ones. Reporting systems were designed precisely to capture this information, and they have always tried to solve the problems of confidentiality and responsibility, with different success rates.

The system learns from events over time, but an important change of emphasis took place long ago: Any field, while it is populated with experimenters and mavericks at different levels, is supposed to produce accidents. The system tries to learn from them, and new events are expected to happen because of experiments or technological adventurism.

However, this only happens in the early development phases, where the stakes are lower than in a developed field. In Aviation, things changed once common people became users, the magnitude of the activity grew and crossing an ocean became for many users an action so trivial as a long bus ride without intermediate stops. Events are no longer allowed, and the emphasis has changed. Learning from accidents has became unacceptable; instead, the system tries to avoid them.

Of course, nobody can deny the appropriateness of this change: It is not acceptable that it should require three crashes, each with more than 500 people on board, to refine the A380 or B747 into a perfect plane. The plane must leave the plant as a product that is good enough to avoid accidents and, if possible, as the A340 did, cease production without a single casualty coming from a flight accident.

A single event can have far greater impact now, since thousands of planes in service are able to transport hundreds of passengers each. The impact is greater still when common passengers become convinced that flying is a trivial and non-risky activity.

Therefore, the objective related to events has became avoidance *ab initio* rather than avoiding repetition. This change, as we will see later, has a great impact on learning capacity.

The Search for Failure

A learning model centered on the correction of identified failures is coherent in an environment where many factors exist whose destructive potential is enough to cause a serious accident without other concurrent factors.

In early Commercial Aviation, many factors existed with that potential. Then, a *post hoc* analysis was conceptually easy but instrumentally difficult due to the lack of resources to store information, even in situations of high-level destruction. The question to answer was "What failed?", knowing that there were many opportunities to find a simple answer, as far as the information could be retrieved.

Then, the main problem was how to reconstruct an event – that is, how to reproduce the facts before the accident beyond supposition. That could make the investigation of an accident difficult, even when there were few factors involved.

> Accidents as shocking as the mid-air explosions of the De Havilland Comet could have been easily clarified, had FDR and CVR existed at that time.

> Another major accident, known as the "Staines case" would be fully clear, instead of being based on suppositions coming from autopsies and the positions of levers as they were found. Actually, the scarcity of available resources with which to investigate this accident would be one of the drivers to developing technical resources aimed at providing the missing information.

Once the environment became more complex, a second model appeared: This model was based on the fact that no single factor could cause an accident by itself. Of course, a mid-air explosion, as with the Comet planes, could cause an accident without any other contributory factors. However, an explosion should not happen because of a single factor but because of an improbable conjunction of factors. This approach led to a different model, centered on breaking causal chains.

Breaking Causal Chains

Currently, accidents are produced as the result of complex interactions between different variables. James Reason popularized this approach as the "Swiss Cheese model". Different barriers would have holes that, if aligned, could lead to an accident.

The virtual inexistence of single causes and the complexity and diversity of causal chains reduce the probability of the repetition of a specific sequence of events into something very remote. For this reason, a safety approach aimed specifically at avoiding the repetition of a specific event should not be enough.

Therefore, the situation changed: It is no longer about preventing the repetition of a sequence of events whose probability is extremely low. Instead, investigation tries to break complex causal chains that could lead to different events, not only those that have already happened.

By the same token, a final event can have very different causes behind it. For instance, runway collisions have happened in the past and we can expect them to keep happening, but the set of "holes" can be different in every situation since different processes could lead to the same outcome.

> The biggest accident of the history of Commercial Aviation happened at Los Rodeos Airport, Tenerife. Two B747s loaded with passengers collided on the runway.
>
> A few years later, another collision on the runway happened at Madrid Airport. However, the processes leading to both accidents were so different that the label "collision on the runway" would not offer any clue as to the means to prevent it from happening in the future.

Instead of addressing the runway collision, root causes are identified in the process. It should be expected that these root causes could prevent some – though not necessarily all – runway collisions and, perhaps, help us to avoid other different events.

An issue identified in the Los Rodeos and Staines accidents was the so-called authority gradient. If we accept that this factor is one of the causes for both accidents, we should accept that it could be behind other very different accidents.

At the same time, even the theoretically perfect management of this issue would not be enough to prevent all runway collisions (Los Rodeos) nor all stalls after take-off (Staines).

Of course, we can still find events that are very specific, repeated and which have the same root causes as the original event – for example, Turkish Airlines 981 or Spanair JKK5022.

Turkish Airlines 981 happened after a close call with American Airlines 96, where a design flaw had been identified but not corrected. The Turkish Airlines case would show not only the improper research of the former case but deeper organizational problems with both regulator and manufacturer.

The Spanair case had a twin brother in Detroit 21 years earlier – an accident where the same system failed. Since this system is included in the minimum equipment list (MEL), something failed in the investigation, allowing a repetition of the event.

These cases, as well as others widely known as the *Challenger* and *Columbia* space shuttle accidents, seriously question the organizational learning model and show that something is wrong there.

In every one of these cases, we can affirm that the analysis of the original event was poorly performed: The failure may have been in the identification of the facts leading up to the event, but the most likely option would be in the assessment of their importance and, hence, in the adequacy of the actions to prevent a new occurrence. Organizational factors and related pressures had a prominent role in the repetition.

We should deal, then, with the first analysis as an event in itself: Discovering why the organization was unable to learn from the first investigation and, hence, why that investigation became useless should be the starting point.

The general objective – breaking causal chains, as a wider objective than avoiding repetition – has led to the appearance of systemic models of investigation. These models try to manage two facts:

1. In complex systems, unforeseen interdependences can appear because of the failure of a single part in the system; for example, a failed sensor can trigger an automatic process leading to a confusing situation. Furthermore, even if every part of the system behaves as expected in the design, they could lead to a negative event in unexpected situations.
2. Those who operate or manage the system are probably unable, due to training and specialized tasks, to anticipate, perceive or even diagnose the interdependence before it triggers an event.

The systemic concept, trying to go beyond the immediate cause, gave rise to different investigation methods that try to identify and fix underlying causes. Widely known models such as threat and error management (TEM), bow-tie risk analysis or the Human Factors Analysis and Classification System (HFACS) are good examples of the concept.

Unfortunately, as some failed investigations show, the investigation does not always reach the underlying causes, concluding with the behavior of the operator closest to the event.

A model such as the aforementioned HFACS can show how this happens graphically (Figure 2.1).

The model has five different levels, and the investigation can stop at any of them, once it has been established that going beyond will not add any valuable information. If the investigation stops at the "Unsafe Acts" level, the conclusion will always point to different modalities of human error. That would leave untouched the levels below it.

A high-level observation of the model will show that these levels are related to different management steps. That should be a strong incentive in

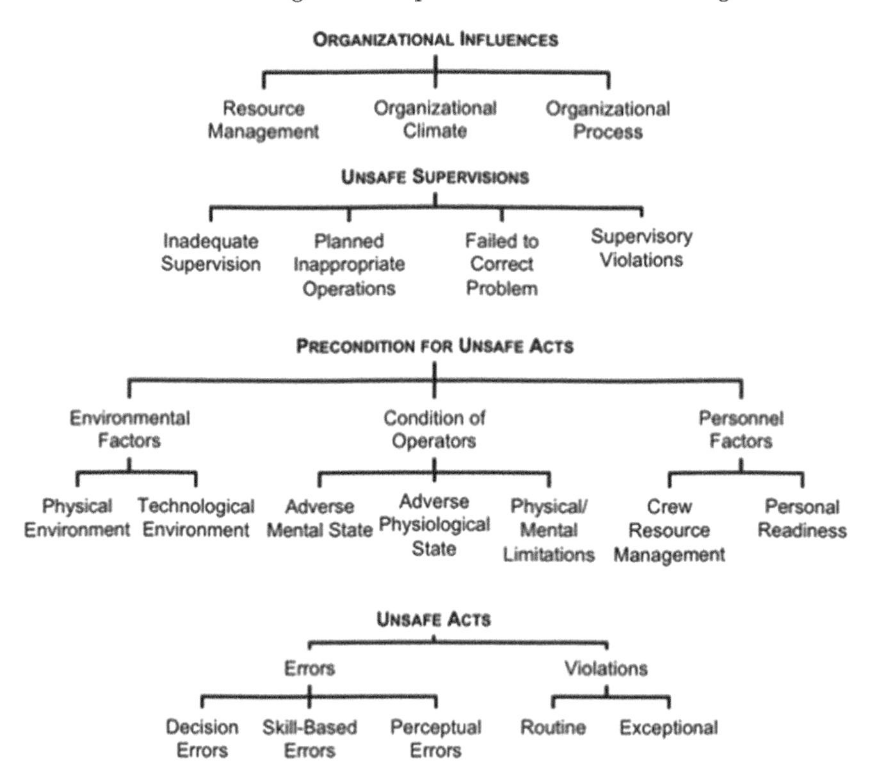

FIGURE 2.1
The Human Factors Analysis and Classification System (HFACS). (From Shappell, S. A. & Wiegmann, D. A., The Human Factors Analysis and Classification System – HFACS, 2000. Retrieved from https://commons.erau.edu/publication/737.)

many organizations to stop at the "Unsafe Acts" level and, hence, that could explain why many accidents are commonly assigned to human operators.

If the analysis goes beyond the first obvious level, different conclusions could be obtained from the same facts. Then, different models can guarantee the breakage of causal chains, but an inadequate decision to stop the process at a specific level can spoil the potential performance of the model. Interestingly, the stopping rule has always been a major problem regarding the development of artificial intelligence. When the information is processed by people, a similar problem appears under a different light: It's not a matter of an iterative, endless process but of a process that, beyond a specific point, does not produce a useful outcome for the analyzed problem.

Stopping the analysis at a higher level than required will then lead to trivial conclusions and no serious changes should be expected. On the other hand, going deeper than required could turn an investigation into a philosophical exercise. Which, then, is the right level of analysis?

The repetition of accidents sharing the same organizational roots shows that the analysis was not performed at the required level. At the same time, some cases can be found where a single accident led to major organizational changes. Examples worth mentioning in this regard are Air Ontario 1363, RAF Nimrod and Mount Erebus. These three cases can be considered as masterworks in the Aviation investigation genre.

In which case, this is not only a matter of available methods or resources; it's a matter of choosing the right focus or organizational links to be analyzed. This is especially clear in those cases showing, through repetition, that a poor analysis was performed the first time.

Establishment of Legal Responsibility

The focus of events analysis has changed due to the evolution of Aviation and its new requirements, bigger airplanes, crowded airspace – all of which have raised the potential severity of a single event.

A disaster because of the existence of a design or maintenance failure in the airplane, an incorrect instruction from an air traffic controller or one that is incorrectly understood by a pilot, or an incorrect or poorly performed decision in any phase of the flight can lead to significant compensation.

Therefore, it is necessary to establish who was responsible for the damages caused. This objective requires, as well as improvement, a careful reconstruction of the facts. A milestone in the assignment of responsibility is the reconstruction of Egypt Air 990:

> A Boeing 767 crashed into the Atlantic Ocean, for reasons then unknown. The analysis of the flight recordings and the remains led to opposite statements.
>
> Egyptian analysts insisted on a technical failure: Another Boeing 767, manufactured at the same time and operated by Lauda Air, was fully

destroyed in a previous accident due to the uncommanded functioning of the power reversal mechanism[1] in one of the two engines.

However, the investigations uncovered a set of facts that left no doubt about the voluntary origin of the event.

1. One of the pilots was alone in the cockpit just before the airplane began an almost vertical descent.
2. When the captain re-entered the cockpit and tried to recover the normal flight position, the actions of the other pilot were contrary to recovery (a fact confirmed by the position in which the rudders were found). Finally, the pilot cut off the fuel to the engines.

The potential impact on passengers of the existence of suicide pilots in the airline led Egyptian authorities to attempt to argue a technical failure (a position that can be found in the official report itself). However, the objection is easily refutable by the data from the flight recorders.

Egyptian authorities claimed a technical failure based on the fact that another plane, destroyed in an accident, had been manufactured at the same time and installed with the same technical fault. At the same time, they dismissed the fact that the pilot was praying "In God I trust", because it was a common act among faithful Muslims. However, a fact present in the voice recorder was omitted: The pilot said "In God I trust" 11 times in a row, the last of them immediately after cutting the fuel to the engines.

The credibility of Egyptian authorities in this issue can be questioned through the analysis of the full available data, not a selected sample. However, there is something to be praised: As shown, they had the opportunity to include in an official report a different view and the causes supporting that view. Many other cases where different parts show different views and explain them can be found.

More recently (2015), the Germanwings 9525 event had a similar origin, even facilitated by the installation of a security door in the cockpit after the September 11 attacks. The difference between both cases is that, in the second one, nobody denied the suicide. The establishment of legal responsibility is alien to the learning process. Furthermore, in some situations, it can be plainly opposite.

First, the potential informants of an event could, at the same time, have contributed via their actions to the situation. If so, the incentive to testify against themselves is scarce. Then, it is important to have the resources to reconstruct an event where the cooperation of the involved people could be dubious.

The existence of penalties linked to responsibility implies a growing pressure toward regulation compliance. Usually, people are considered responsible if there is a set of regulations that can be applied to the event. Then, the actors find themselves limited by the regulations in terms of their role and the control they have on the event.

That should not be negative but, due to the growing complexity of the system, the regulations can find "unforeseen" situations that were, apparently, foreseen. Many situations could be shown to be affected by this factor. Once someone has chosen an option, it can be difficult to find out what consequences would have come from a different decision.

An example and a counter-example are as follows:

- The example: The only accident on a Concorde happened after the pilot, following the rules, took off with a major problem, which appeared when the plane was already going faster than the decision speed. The outcome was the destruction of the plane and the death of all the occupants and several people on the ground.

  What if the pilot had decided not to follow the rules and remain on the ground? Once this decision is made, nobody could say that the alternative decision would mean a full disaster. Had the pilot survived, we could expect him to be considered responsible for the accident.
- The counter-example: On Spantax 995, a heavy vibration appeared after decision speed. The pilot was convinced that the plane could not fly and stopped it, conscious that there was no room to stop the plane. The plane was destroyed by fire and some of the passengers were killed, many of them while going back to the already evacuated plane before the fire.

  The subsequent investigation revealed that the problem came from the nose gear; the plane could fly, and the vibration should have disappeared once airborne and with the gear stopped.

  The problem: The extreme vibration was an unknown phenomenon for the pilot. He could not know if it was due to the wheels, the flight controls or a structural problem, as established by the official report:

  > "The decision of aborting the take-off, though not in accordance with the standard operation procedures, is in this case considered reasonable, on the base of the irregular circumstances that the crew had to face, the short period of time available to take the decision, the lack of training in case of wheel failure and the absence of take-off procedures when failure other than that of the engines occurs."

A similar situation happened more recently in the case of US1549. Full loss of power was foreseen above 20,000 ft. but not immediately after take-off. The pilots were supposed to use a long checklist and, after that, land at an airport. The facts showed that this was impossible, even when a simulated flight showed that landing had been possible *if* the pilots had tried to land immediately after the loss of power.

The USAir and Spantax cases found something positive regarding the responsibility of the pilots: In the first case, it was possible to see – through simulation – the consequences of the alternative decision. In the second case, the reasons for the take-off rejection could be easily understood.

The Concorde case is more complex: Had the pilot decided to remain on the ground, could anyone affirm that the plane would not have been able – as the facts showed – to perform an emergency landing after taking off?

We know now, as the facts have clearly showed, that this option did not exist, but that would have been hard to establish if the decision had been the opposite. That can make things very difficult regarding legal responsibilities.

Investigations of events in the Commercial Aviation field and those done by the Justice Department have different objectives. The first is aimed at learning and increasing safety, an objective explicit in many accidents reports, a fact established by ICAO Annex 13. The second investigation tries instead to set responsibilities.

These different objectives can lead to conflicts, which are solved in different ways by the authorities. Inside the Aviation field, accident investigations are – at least theoretically – performed by organizations separate and independent from the Aviation Authority.

The same principle applies to reporting systems. A good example of independent use can be found in the United States: the ASRS reporting system is not dependent on FAA (the Aviation regulator) but on NASA, a powerful organization not subordinate to or subject to pressure from FAA, and establishes a non-punitive approach.

Many events, then, are reported by people that could be punished because of the facts reported. In which case, a strong incentive to report is precisely the avoidance of eventual penalties.

Within the Aviation field, things are solved or on the way to being solved regarding the use of information and its effects. However, things are different when data from an accident investigation are claimed from the Justice Department. ICAO establishes in its Annex 13 some recommendations related to the data used in investigations. For instance, it points out that the records of investigations will be used only for investigative purposes but, at the same time, an authority designated by the state may ask for these records.

In plain words, that means that, in many countries, a judge can ask for all the documents handled by the Investigation Commission. At the same time, the decision of that judge is not constrained in any way by the conclusions from that commission, whose role is considered to be that of a technical advisor. Certainly, a sentence contradicting the conclusions of an official report would be easy to attack but, anyway, it is easy to appreciate that different dysfunctions exist in the relationship between both environments.

As a side effect of these dysfunctions, some organizations will stop their internal investigations – aimed at improvement – if they foresee that an event could have consequences leading to a criminal investigation, and they could be asked for all the data coming from the investigation.

Beyond that, in some situations a destructive test could be required, but it could be stopped by a judge if the element to be destroyed must be kept intact as evidence in a hearing.

This is a conflict that, with differences among countries, can be considered unsolved. When the consequences of an event do not require a criminal investigation, the Aviation field tries to guarantee data confidentiality and its correct use, and, with different development levels, a good solution can be found in many countries.

On the other hand, if consequences lead to a criminal investigation, confidentiality can be broken. Hence, the information will not flow openly in major events.

In summary, there are good resources of information, both on the technical side (flight data recorders, cockpit voice recorders, air traffic control records, simulators, etc.) and on the social side (reporting systems, non-punitive policies, etc.) but, in some situations, the system can fail on the incentive side.

Event-Based Learning Cycle

As already pointed out, damages produced in the most serious events – and hence, those requiring an explanation with the utmost urgency – have forced the development of recording devices, channels of information and the wide use of statistical information to make decisions about risk.

Different phases can be used to define the life cycle of the information about an event. In the first place, the gathering of information in serious events will be based on installed recording devices, while in minor events the cooperation of the actors will be required to identify potentially dangerous situations.

After that, the relevant information must be sent to those actors interested in the causes of an event that they could suffer. Finally, the event and the subsequent analysis can lead to decisions aimed at avoiding a future similar event or, if that is not possible, decreasing the impact. Each of the phases of the information life cycle are as follows.

Information-Gathering Phase

The analysis of an event is only feasible if the information on the facts leading to it is sufficient and reliable.

As such, devices and operational procedures have been developed whose specific goal is to permit the reconstruction of the actions prior to an event. In this way, its causes can be determined. Today, unexplained events in Commercial Aviation are extremely uncommon.

Commonly, accident reports include recording transcriptions, comprising conversations in the cockpit and radio and data transmissions. When required, they will include recordings of engine power parameters, position, and so on.

Outside the plane, the legal situation of the crew, training, activity records, maintenance records, registration and varied data about the plane and any identified weaknesses in the design are included.

Finally, the physical analysis of the plane after the event, together with all the aforementioned information, usually leaves little margin for doubt.

The recording devices on board an airplane, informally known as "black boxes", have two components: The first, the flight data recorder (FDR), records the heading, speed, flight levels, vertical acceleration and microphone use, among other things. The second, compulsory in airplanes since 1966, is the cabin voice recorder (CVR), which weighs about 9 kg, records all sounds in the cockpit, can withstand an impact with a pressure of 30,000 kg and has a notable resistance to fire.

The gathering of information has permitted its use, directly or through the introduction of the known parameters in a simulator. The observation of the behavior of the simulator can often provide additional information.

Additionally, the gathering of data for its conversion into statistical information also forms part of the same process, even if in this case it is more directed at making risk-based decisions than a causal analysis of a specific event.

Decisions regarding the redundancy of vital devices or the authorization to use a device are taken in respect of probability calculations, which in turn are constructed on information derived from events.

Three cases of the use of statistics for making risk-based decisions are as follows:

- The first case is illustrated by the DC10 and known as U232. In the DC10, three independent hydraulic systems were installed under the assumption that the probability of three systems failing at the same time was below the probability admissible for catastrophic risk (1 in 1 billion).

  In this case, the calculation was performed during the design, and the information from this event would lead to an adjustment of the estimation. The U232 case, which will be widely commented on, showed that the three systems could fail at the same time and forced a radical modification of the assumption.

- The second case consists of the authorization for the transoceanic flight of twin-engine airplanes. This authorization is based on the requirement for technical measures and statistical recordings that check the real reliability of their engines. With this objective in mind, even small deviations from normality are recorded to detect in advance the tendency for the failure of a specific engine.

  Nowadays, a vast majority of high-capacity airplanes are equipped with only two engines and certified to make transoceanic flights; an engine failure over ocean means the airplane should remain in flight with only one engine during a period that can exceed six hours and with many people on board. Furthermore, that time has been growing based on statistical analysis.

This type of decision depends on a constant gathering of data, which, eventually, could require their revision. Curiously, as previously mentioned, the system can be a victim of its own success, since due to the reliability achieved, it is difficult to assemble enough cases to make a valid sample and, hence, to determine the real reliability of a single engine subject to the extra effort required to maintain a twin-engine airplane in flight.

Faced with this fact, the certifying authorities require that in these circumstances – a twin plane powered by a single engine due to the failure of the other – the working engine should not exceed its normal operating parameters for pressures and temperatures; that is, the engine should not exceed the value established for maximum continuous thrust (MCT).

- The practice called flight data monitoring (FDM), similar to ETOPS records but extended to systems other than the engine, is based in the use of recording devices to store statistical and, unlike ETOPS, misidentified information. Frequent problems in a specific fleet can be identified by recording minor events and, hence, trends.

Together with recording devices and the gathering of data for statistical processing, voluntary reports play a major role in information gathering. Recording devices cannot avoid two problems: Their analyses are laborious and, hence, they are only performed when the importance of an event requires a reconstruction. Both recording devices and reporting systems supply a medium for storage, but they are useless without analysis. Furthermore, recording systems are automatic, but reporting systems must include in their organizational design a good motive for their use.

Different situations, not recorded as anomalies by the systems of the plane, with potentially serious consequences would go unnoticed and would not generate learning without a voluntary action on the part of those involved. Many relevant cases could be lost without the active involvement of the people who report them.

Information on these events is a valuable instrument, but there are three barriers, two of them clearly identified and usually managed. The third one is less well known, and it could be especially relevant in the coming years.

- Existence of penalties: An informant, if involved in the situation, could hide information.

  The importance of the problem has been acknowledged by the main regulators, and the common solution is giving preference to improvement over responsibility. That can be done as far as an event does not have consequences leading to criminal liabilities.

Of course, this practice leads to a certain amount of fraud – that is, people reporting to avoid penalties. However, that price is still low compared with the consequences of hiding information.

- Confidentiality: There are many reasons for a reporter to be anonymous. Any reporting system should guarantee that point but, even if formally guaranteed, the information in the report could be enough in some cases to identify the reporter. If that happens, or people suspect it could happen, important information will be lost, especially if that information points to managers that could take revenge on the informant, if identified.
- Apparently trivial issues: The motivation to report decreases with the perceived importance of the issue to be reported and with the length and complexity of the reporting process. Very frequent failures become a part of the environment and they are not reported, especially if there is not an answer to the report or things are not modified.

The problem with unreported "trivial" issues is that some major decisions could be taken with the wrong information. The statement that 80% of accidents come from human error is commonplace, but this statement, true or false, is not the whole picture. To get the real picture, we should answer two questions:

1. How many accidents are avoided through human intervention?
2. How many human errors come from confusing design?

Human intervention does not need to be heroic, exceptional or, visibly, a death-or-life issue. Trivial interventions that, as such, are unreported, could stop in an early phase a process leading to a major event.

Furthermore, interface-oriented designs can display an apparently easy-to-handle system whose internal complexity is fully unknown to the user. That internal complexity can lead to confusion and related mistakes. Some examples of this situation have already been shown.

The "80% of accidents come from human error" mantra leads people to ignore the former two points, and some decisions could be taken based on incorrect assumptions. For instance, some Aviation managers have begun to speak openly about the possibility of removing one of the pilots from the cockpit.

If something like that is seriously evaluated, dramatic events such as incapacitation, heart attacks and so on will be kept in mind but, for instance, minor disagreements, discussions,[2] lapses captured by the other pilot and many other micro-events, usually unreported, would not enter among the parameters for the decision.

Of course, this is not a problem affecting only pilots. A fair amount of automation issues written by controllers can be found in the ASRS reporting system and, in every single case, the passengers, after leaving the plane, were convinced they had been on an ordinary flight. Actually, looking only at the outcome, they were. However, converting the flight into an ordinary flight very often requires minor non-standard actions. It is not only an issue of avoiding human error. If the flight became ordinary under a non-standard situation, it's because the human positive and invisible side of the system went far beyond compliance with the standard operating procedures (SOPs).

So, although the resources – technical and non-technical – for acquiring information have clearly improved, other things should still be improved, mainly those that end in a non-noticeable outcome, whatever happened on the way. As a small sample, without pretending to have statistical validity but only to explain the phenomenon, some cases taken from the Aviation Safety Reporting System (ASRS) database (2019) are as follows:

- B767-300 First Officer reported receiving an apparent false RA on approach to SEA; afterwards, ATC reassured them of no reported traffic in the area.
- Air carrier Captain reported receiving a GPWS terrain warning on approaching to BHM that could have been a false warning or possibly a high descent rate toward terrain.
- Air carrier flight crew reported an EGPWS "TERRAIN, PULL UP" warning while flying a left descending base for a visual approach to TYS Runway 23 L.
- B787 flight crew reported the gear down overspeed warning was being left in "Gear Down" position during a maintenance action, so after takeoff the crew was not able to accelerate or climb to a normal altitude. It was a false warning, so after consulting with ground support they were able to continue and land normally.
- B737-800 First Officer reported receiving a false traffic alert at FL360 with no traffic in the area.
- Airbus A321 flight crew reported receiving an aural alarm and an "AFT CARGO FIRE" ECAM alert which later became a "SMOKE AFT CARGO DET FAULT". The crew elected to divert and land as a precaution. There was confusion about the indications and checklist procedures as there is no procedure for "CARGO FIRE", only for "CARGO SMOKE".
- An A321 flight crew landing in VMC on Runway 28 L at FLL received a GPWS Terrain warning as they crossed the threshold, closing the throttles for landing. The reporter suspects the 55 foot raised elevation at the approach end of the runway and the "seawall" type structure supporting it contributed to a false warning.
- As they began their approach the MD-82 flight crew received a DR (dead reckoning) warning they switched from GFMS guidance to raw data. After established inbound with the ILS and VASI centered and the runway in sight they received an EGPWS TERRAIN warning which they silenced and proceeded to a safe landing. At the time

of the warning the NAV display showed them well to the right of the runway.

- B737-800 Captain reported intermittent false engine fire warning in cruise flight. After consulting with ground personnel the flight continued to destination.
- B737-700 flight crew reported receiving false wind shear warnings on two approaches to Runway 27 at BOS, commenting this is a known problem.
- B737 crew has a TCAS RA at 1,200 FT after takeoff from SAN and attempts to comply with the resolution advisory. The target remains co-altitude and ATC advises there is no traffic just before the GPWS advises "don't sink." TCAS warnings are then disregarded and cease altogether at 3,000 FT.
- A B757-200's R AFT EMER DOOR EICAS alerted after takeoff but the crew determined after completing the QRH that it was a false warning and with the cabin fully pressurized continued the filed destination.
- Embraer Legacy 550 Captain reported the simultaneous failure of both Flight Control Computers (FCC), and both Attitude & Heading Reference System (AHRS) computers immediately after takeoff, resulting in the aircraft reverting to Direct Mode for operating the flight controls, causing the aircraft to divert to a suitable airport.
- MD-11 flight crew reported failure of Central Air Data Computer at cruise in IMC conditions.

Many more cases can be found, especially if ATC and airport handling issues are included, where an ordinary flight, despite the outcome, had been anything but ordinary.

The preceding cases each show an important piece of commonly lost information that is hence disregarded when decisions are taken about the human role in the system, the subject of the following chapters.

Very often, an automation failure, a malfunctioning sensor or a problem in the controls are checked using resources external to the system – for instance, the senses of the people involved. A pilot can decide that a ground alert is false for different reasons:

1. The visibility is good, and the position can be visually checked.
2. By cross-checking information from other sources.
3. By knowing that this is a common failure in that plane.

Whatever the option, the pilot won't accept at face value any information that the system might supply. Hence, the human decision in these cases can be very different from – and usually better than – the expected outcome from highly automated systems.

Events that might occur due to incorrect input, followed by overreaction or a standard answer, usually don't. Since human intervention avoids abnormal outcomes, the whole process remains disregarded, hidden or, at least, downgraded. Hence, abnormal processes that end with an ordinary outcome are out of the mind of the designers. Key decisions, then, about how the system should work can ignore these highly informative cases about the human role.

In summary, it could be said that information gathering has a major flaw: It's designed to collect all the information from events but, at the same time, it loses relevant information about non-events – that is, situations where the outcome was not noticeable but the process leading to it was. The preceding examples emphasize the importance of using that information in decisions about the future design of a system.

Information Distribution Phase

As already mentioned, information gathering is critical, and some situations can be found where information is still missing, despite the growth of technical resources devoted to this.

The second problem is about what to do with the available information. It must be defined who has unrestricted access to the information and who doesn't, and why.

It must be highlighted that the information that flows inside the system on potential or real events is much broader than the information accessible to users. This is a mixed blessing.

1. It's positive because it avoids creating alarm as a competitive practice. Additionally, all those involved can benefit from the experience of others through the flow of information among them.
2. It's negative because, by not competing in safety, there could be effective reductions in the level of safety by agreement among operators. As already said, the balance between risks and benefits could be socially acceptable if reached through a public process but unacceptable if reached through agreements inside an inner circle, fully opaque to users.

However, it must be noted that there is no shared practice regarding access to information.

A good example is the ASRS practice: Any person with internet access can search the database. Furthermore, it is possible to download it to search with more powerful tools than those available on the website.

The user is not questioned about the motive for their interest nor required to set up a user account. The information that can be obtained in many issues enables even statistical analyses. The system can be of big help in the investigation of accidents, not only by showing precedents but by helping to evaluate the frequency and the importance of a fact – for example, setting the wrong configuration.

Another organization, the National Transportation Safety Board (NTSB), also has a searchable database – again public and with no questions asked – albeit more limited in size, since the cases included are those important enough to require a full investigation.

These excellent practices are far from being commonplace. Many other systems can be found without a searchable database, with less valuable information due to problems in the information-gathering phase or, plainly, restricted for anyone who is not explicitly authorized.

Instead of getting raw and easy-to-access data, people outside the system usually receive sweetened statements about how safe and how perfect the system is, while hiding some other issues. For instance, a manager of courses for people with a fear of flying affirmed that a large airplane with its four engines stopped could fly 200 km from the cruise level before reaching the ground.

That is approximately true, and cases such as Air Transat 236 might certify the point. However, the affirmation gives a false image through a half-truth: While it's true that the gliding capacity of a large airplane with the engines shut off would allow it to cover a long distance, it suggests that a full loss of power would be a trivial incident.

Some technical debates hardly reach the world outside Aviation practitioners before reaching an agreed sweetened version. Everyone seems to assume the obligation of offering an "everything is under control" picture and there's nothing to worry about but price and comfort.

The good part of the opacity to outsiders is the possibility of allowing the information to flow inside the system without obstacles, with the consequent speed to set up corrective measures. The bad part is the potential for the public to suffer a nasty surprise, leading them to distrust the whole system.

Mauriño et al. (1995), quoting the analysis of the accident of the space shuttle *Challenger*, extract the following paragraph from the conclusions:

> All organizations are, to varying degrees, self-bounded communities. Physical structure, reinforced by norms and laws protecting privacy, insulates them from other organizations in the environment. The nature of transactions further protects them from outsiders by releasing only selected bits of information in complex and difficult-to-monitor forms. Thus, although organizations engage in exchange with others, they retain elements of autonomy that mask organizational behaviour.

The first obvious point to highlight is that this was written before the 2003 *Columbia* disaster. Then, a new disaster would show in a dramatic way that an accurate description of the environment is not enough, if it is not followed by suitable actions to change that environment.

As a generic description, the idea of a boundary between organizations could be proven valid and, as the facts would show later, that is difficult to change. However, the description of the environment would be valid for the Commercial Aviation field. A problem could still be added that was not explicit in the analysis of the *Challenger* event, even though it was present.

It is not only about a main organization surrounded by subcontractors, but about organizations competing among them. The *Challenger* case is not so different from Aviation dynamics since those subcontractors are competing among them, as Aviation operators and manufacturers do.

The behavior of these subcontractors during the development of the project is going to define the chances that they will have to obtain the next assignment. Otherwise, the behavior of Thiokol during the *Challenger* event, changing from their correct initial assessment to a false no-risk situation, would be hard to explain.

The idea of boundaries among organizations is correct, but the reality goes far beyond that: These boundaries are strictly defended, and the behavior of the organizations is conditioned by that defense and the resulting pressure from the top.

Despite these fights among the organizations involved, the different stakeholders in the inner circle behave like a block about decisions regarding which information should be disclosed or not. This practice can be beneficial – there is a limited war with clear boundaries – but it brings its own risk. That risk is not only about material and life losses; it is also about public acceptance and its consequences.

Information Utilization Phase: Generation of New Abilities

Once the cause of an accident is identified, the immediate action is corrective in character if the available technology exists or preventive if the event is one that cannot be confronted successfully.

Very often, both types of action are linked, and technology is used as a form of prevention. One example is the installation of meteorological radar aboard airplanes. The radar does not make storms less dangerous, but it facilitates the avoidance of its most active zones.

Nevertheless, since practically all accidents are multicausal, the learning process advances more through the identification of critical points than through the avoidance of a specific event. Obviously, at the same time, if an event identical to a former one happens, it would be a clear proof of the false identification of critical points.

As an example, the investigation and subsequent recommendations after the Los Rodeos accident is representative of this idea. The magnitude of the accident, which involved the collision of two Boeing 747 airplanes while one of them was attempting to take off, could invite radical changes throughout the whole system. Instead, trying to break causal chains, the following recommendations were made in the official report (Comisión de Investigación de Accidentes e Incidentes de Aviación Civil, 1977):

1. Emphasis on the importance of exact compliance with instructions and clearances.
2. Use of standard, concise and unequivocal aeronautical language.

3. Avoidance of the word "take-off" in the ATC clearance and adequate time separation between the ATC clearance and the take-off clearance.

This accident can be used as an extreme example of the concatenation of circumstances that can happen in a serious accident. Many minor or apparently unrelated – but non-trivial – facts were required to have that outcome. The absence of only one of them would have averted the accident. This fact, on the other hand, can validate actions that, after analysis, appear weak, especially compared with the magnitude of the accident.

1. Closure of Gran Canaria Airport due to the explosion of a bomb and the threat of a second.
2. Marginal meteorological conditions at Tenerife Airport.
3. Obstruction of the exit taxiway on the ramp (congested due to the closure of Las Palmas) by the KLM airplane. The PAN AM airplane was ready for take-off earlier, but it had to wait because it did not have enough room to exit. That provoked the temporal coincidence.
4. Great difference in status between the captain and the first officer in the KLM, meaning the latter failed to keep a vigilant attitude over the actions of the captain.
5. Flight time limitations: Crewmembers were near to the legal limits. Therefore, they required a fast take-off or a cancellation, with all the associated problems.
6. Mistaken intersection by the PAN AM airplane when leaving the active runway due to visibility and the orientation of the exit.
7. Temporal coincidence between the radio transmissions of the tower and the PAN AM airplane interfering with each other and being poorly received by the KLM airplane. The content of the transmissions ("Wait and I will advise" from the tower and "We are still on the runway" from the PAN AM airplane) suggests that the reception of any of them would have indicated that the take-off should have been immediately cancelled.
8. Use of a reduced-power take-off technique, which, in exchange for reduced fuel consumption and extending engine life, lengthens the take-off run of the airplane. The importance of this factor can be appreciated knowing that the first impact took place between the landing gear of one airplane (KLM) and an engine of the other (PAN AM). Put another way, when the KLM plane crashed against the other plane in the runway, it was already airborne.
9. Use of a non-standard phraseology, not understood by either the control tower or the captain of the PAN AM airplane, regarding

what they were doing. For example, "We are now at take-off" could be interpreted both as "We are now at the take-off point" or "We are now taking off."

10. Lack of recent practice on the part of the KLM captain as pilot-in-command, having lost familiarity with the type of incidents that can arise and increasing the levels of anxiety associated with such incidents.

Similar conclusions with respect to the interaction between circumstances and the need to find a critical factor could be obtained from case TWA-800 in which a TWA Boeing 747 exploded in flight shortly after take-off. According to the official report,[3] the circumstances that led to the accident are as follows:

1. A hot day and a delay in the flight led to the air-conditioning being kept running for a long time.
2. The engine for the air-conditioning system was close to a fuel tank and heated its interior.
3. The fuel tank was empty. Nevertheless, the gases, present in the tank because of its regular use, became explosive at a lower temperature than liquids. Therefore, an empty fuel tank represented a greater risk of explosion than a full one.
4. A short-circuit in an instrument provoked an electrical spark, which ignited the explosive atmosphere in the fuel tank, producing the explosion of the airplane.

Unlike the former case, there was not a major operational failure here but a design failure that was exposed through a set of exceptional circumstances that did not appear in more than 25 years of operation of the airplane type. Therefore, the only action possible would be design change.

In this way, we attempt to determine which actions can achieve better results in the task of preventing the future occurrence of similar events.

Hale et al. (1997) point out that the analysis and learning process generated models, tools and estimates about which there existed some consensus for the first two "ages" of safety, centered on avoiding technological and human failure, respectively.

A third age, in which we might be right now, has a different concern: complex sociotechnical systems and safety management. Both analyzed cases show how the convergence of small causes can lead to a major event.

This situation justifies taking into consideration a systemic principle: Cause and effect can be in a very different magnitude range.

The requirement to bear in mind the interaction between variables isn't the only problem that appears before we retrieve usable information from event analyses.

Another complication comes from the moment of the analysis and the resulting frame of mind. Einhorn and Hogarth distinguished between *thinking backward* and *thinking forward*:

- Thinking backward, a basic task in the analysis of an event, is defined as a primordially intuitive task, diagnostic in character and which requires capacity of judgment in the search for clues and the establishment of apparently unconnected links, such as the test of possible causal chains or the search for analogies that might be of help.
- Thinking forward, a task performed when attempting to prevent an event, is different: It does not depend on intuition but on a mathematical proposal. Whoever needs to make the decision should group and weigh up a series of variables and then make a forecast.

However, with a strategy or with a rule, the reliability and accuracy of each factor must be evaluated and the person making the decision reaches an integrated prediction by combining elements.

Limitations of Event-Based Learning

Following these two modes, thinking backward and thinking forward, we could say that event analyses are performed under the modality of thinking backward. However, once conclusions are obtained, any resulting process to be implemented is carried out under the modality of thinking forward.

Both modalities share a very strong link: The solutions are adopted in full agreement with the conclusions about what happened and why. Then, any bias leading us to ignore some parts or emphasize others will be reflected in the solution.

The relevance of this fact grows with the complexity of the system, where many variables are criss-crossed in different ways, and choosing the right one is far from being a straightforward task. In this sense, Luhmann (1993) suggested the concept of *hypercomplexity* as a limit to learning. Hypercomplexity is defined as "a system state that happens when every part tries to optimize from its specific point of view" – in other words, development without integration. Thinking forward in a hypercomplex environment becomes increasingly difficult. In a similar line, Perrow pointed out that the problem regarding complex organizations is the potential unforeseen interaction between functionally unrelated units as a side-effect of their proximity.

Manufacturers, as well as regulators, try to anticipate these interactions in two different ways, attending to their specificity:

1. Analysis of past events to know what failed and what related fails could happen.

2. Analysis of the risk linked to systems using flammable fluids, high pressure, high voltage, potential corrosion, explosions, unintended damage by people and many others.

One example of the first class can be found in the Safety Recommendation from the NTSB after the TWA-800 event (National Transportation Safety Board, 2000).

> The potential for short circuits to damage nearby wiring (more than 1 1/2 inches away) has been documented in Safety Board investigations of numerous accidents and incidents. The Safety Board concludes that existing standards for wire separation may not provide adequate protection against damage from short circuits. Therefore, the Safety Board believes that the FAA should review the design specifications for aircraft wiring systems of all U.S.-certified aircraft and (1) identify which systems are critical to safety and (2) require revisions, as necessary, to ensure that adequate separation is provided for the wiring related to those critical systems.

Related to this issue, but with a more general view, we can find another example, taken from EASA CS25 AMC Appendix H:

> Determination of EWIS[4] separation requirements is required by 25.1707 … For example, if an aeroplane has a fly-by-wire flight control system and a minimum of 2 inches of physical separation is needed between the EWIS associated with the flight control system and other EWIS, this information should be available in the ICA.
>
> Similarly, the separation of certain wires in fuel tank systems may be critical design configuration control items and therefore qualify as an airworthiness limitation. Maintenance personnel need these guidelines and limitations because many times wire bundles must be moved or removed to perform maintenance.

The preceding examples could be enough to show how an event-based learning process works: An event shows something that was considered unthinkable or was simply unforeseen. Once it has happened, there is first a movement aimed at preventing or mitigating similar events. If the causes justify it, there is a second movement to establish a more general rule.

However, as already mentioned, complexity can lead to new and unforeseen issues. When new issues are solved by adding more complexity, a paradox can appear.

A new accident could take place due to the effort to prevent the last one. This example, from the El Al 1862 case, illustrates the point.

> A failed safety device was behind the El-Al 1862 accident at Schiphol airport in 1992. The first point is that the accident did not happen because the safety device failed and something that should be prevented by that device occurred. The situation is far worse: Had the safety device not been present, the accident would never had happened.

The device was designed as a weak point in the link between the engine and the wing. Its function consisted in ensuring that, in cases of an extreme torsion, the device would break, leading to a clean separation of the engine from the wing. Then, the lost engine would not carry along with it a part of the leading edge, which would make flight impossible.

From the point of view of maintaining the aerodynamic qualities of the airplane, the solution appeared correct. However, as the safety device was defective, it not only failed to perform its intended function, but it would become a major factor in the accident. In this case, the defective device caused the undue separation of an engine, which would hit the leading edge and the other engine on the same wing. Therefore, the accident became unavoidable.

Since this case happened in 1992, we might think that things would have advanced since then (they did) and, hence, we should not find any surprises. In fact, they keep appearing, as the QF32 case shows:

> The QF32 event in 2010 was another "creative interaction" that was hard to foresee. An engine stop was foreseen, but in a four-engine plane it should not have been a major issue, even on an oceanic flight. An uncontained failure that converts an engine into a bomb is much more serious. It did not break the fuselage, but it broke electrical and data wires as well as hydraulic and fuel pipes. The pilots had to manage a situation that had never been foreseen.
>
> It must be added that the fault in the engine design had been identified before this event, but nobody was able to foresee this outcome.

The QF32 event could be used as a model to illustrate the warning from Perrow about increasing complexity: It happened because of an unintended interaction between unrelated systems.

Beck describes these situations as *manufactured uncertainty*. This uncertainty supposes not only an incomplete knowledge base but a surprising fact: The increase in knowledge frequently increases uncertainty too.

So, the mathematical proposal that involves Einhorn and Hogarth's *thinking forward* approach becomes increasingly difficult: The possibility of unintended and unforeseen relationships between variables makes it impossible to consider all of them in the design process.

The concepts of thinking backward and thinking forward imply frames of mind related to anticipation or reaction, something already introduced as being linked to early or mature development phases. However, there is still a third obvious and different thinking model – that is, a model used in operation that, as with the two previous ones, has its own working rules.

Reason (2016) groups his analysis of human abilities under the epigraph "the design of a fallible machine", referring primarily to the limited capacity for attention and, hence, to deductive processes that are also limited, since they operate on the data perceived by that limited capacity.

Under these conditions, human operators find themselves managing a situation based on their belief about how things work. This belief comes partially from their own experience, allowing them to reject options, operating with those considered most probable. However, since not all the necessary data is available, they use heuristic reasoning, far from common algorithmic models.

The relevance of this point is based on a single fact: Events are avoided as much as possible and, hence, they should not be the main learning source. At the same time, forward thinking is difficult in a hypercomplex environment. Then, people could be the only remaining emergency resource in some situations. However, to keep that resource active, we should be aware of how people work and, hence, we should prepare a suitable environment for the features of people.

Of course, that would lead us to enter into a deep discussion about the role and the value of people in a technology-driven environment. While for some, human operators are an unavoidable nuisance to eliminate as soon as technology allows it, for some others, human operators add clear value.

Heuristic reasoning is not always, as pretended by some authors, a "quick and dirty" way to solve a problem. Instead, *frugal heuristics,* as Gigerenzer (2007) calls it, can solve a problem in a different, faster way, requiring far less processing resources than advanced information systems.

Certainly, heuristic reasoning can fail but, at the same time, it can solve problems at speeds unreachable by a computer through rejecting unlikely options, directly or by introducing new parameters in the processing.

> The aforementioned QF32 case shows an example of heuristic processing: After the uncontained failure of an engine (number 2), the systems of the plane sent an unmanageable number of error messages; among them, a failure in the engine number 4. The pilots reasoned that, to reach engine number 4, located at the opposite end of the other wing, the shrapnel coming from the uncontained failure would have to pass through the fuselage, and that, obviously, did not happen. From that moment, the pilots decided to distrust the information coming from the plane systems and try to check everything by themselves.

In the example, then, an unmanageable situation does not become manageable by adding new information. Instead, it became manageable by eliminating items to be processed after a *eureka* moment. Actually, that *eureka* moment is the only new input.

Hypothetically, it should be possible to have a system using resources to apply *think forward* to the present situation and without the limitation of the slow speed of calculation of the human operator.

However, this hypothetical system only links the relevant variables if they were previously identified in its programming. Hence, that system could become useless before an unforeseen situation.

Indeed, an event analysis, less time-constrained than an in-flight event and suitable for a common algorithmic approach, can show this shortcoming in the technological approach.

Flight simulators, as previously mentioned, are used in accident investigations to determine the behavior of the airplane they simulate. Despite the experience acquired in the design of these systems, there are still new situations.

> An event finished with a runway overrun by an Iberia Boeing 747 in Buenos Aires.[5] The airplane suffered an engine failure just before the decision speed. Applying brakes at maximum power resulted in blowing out sixteen tires but, contrary to expectation, the plane did not have room to stop on the runway, exceeding the end by some 60 m.
>
> The use of the flight data recorder to test in a simulator why the airplane did not stop did not clarify the matter. The simulator tests, given the brake pressure and the moment at which it was applied, showed the airplane systematically stopping before the end of the runway.

The solution appeared when the existence of a missing variable was identified: The continuous application of the maximum brake power literally melted the braking devices, which became useless. The simulator used hydraulic pressure data during braking and its corresponding stopping-time calculation, but it did not include the physical deterioration of the device and the added space required to stop the plane. Why? Simply, nobody foresaw that parameter and its effects in the simulator design.

Analysis, backward or forward, is carried out based on mental models coming from a dominant logic. This leads to the inclusion in the analysis of some variables and the exclusion of others. The mental models resulting from this selection of variables can represent a limiting factor for the gathering of results.

Both elements – type of analysis and mental models – are found to be closely related, as normally, analyses are not only carried out at different times but by different persons with their specific mental models.

In this way, design – forward-thinking activity – is performed by engineers with their mental model. The management of the event is handled by pilots, who also have their own mental model, conditioned by the environment. Finally, the investigation –backward-thinking activity – is usually carried out by specialized officials with the technical support of the former and with their own mental model.

Then, as well as the differences imposed by the environment where the processing and the subsequent decisions are made, the different actors impose their own mental model, emphasizing or limiting the contribution of different factors because of their own information structure. A well-known example can illustrate this point.

> If we analyze the official report of the Los Rodeos case – which took place in 1977 and is still the worst accident in the history of Commercial Aviation – it is surprising how a relevant fact is not mentioned: The recent

> flight experience of the KLM captain was related to simulator sessions, where take-off clearances, the logging of passengers on a cancelled flight and flight time limitations did not exist.
>
> As the flight records show, the captain had performed fewer than 260 hours of flight per annum over the past 6 years, less than half the normal amount. Furthermore, in a long-range plane such as a 747, that means a very low number of operations – and the problems pointed out are more related to the number of operations than to hours flown – and, hence, his recent experience at dealing with those problems was still lower.
>
> Despite the stressful situation, the KLM captain performed the checklists by the book. He failed in the least practiced part (take-off clearances) and, probably, administrative problems such as passenger logging or flight time limitations were far more stressful for him than for a common pilot, who would be much more used to such situations in daily life.

We could go still further: If we ask who can show excellent knowledge of technical procedures and, at the same time, fail in something as basic as a take-off clearance, while being so stressed by relatively common issues, a possible answer appears: a trainer, not a common pilot. They can have a common technical base and many common skills but a different mental model that could lead to undesired consequences, if used in the wrong place.

Later, some criticisms would appear about the rarely questioned role of "management pilots". Their current flight practice would be far lower than that filed by other pilots without management positions. According to Beaty (1991), this fact is seen as a natural privilege for those in positions of authority. Therefore, it could happen that the pilot whose recent practice in the fleet is the lowest is also the one who flies the most relevant people or makes the most relevant flights.

These questions were left out of the investigation, revealing also a mental model oriented to hard facts while ignoring relevant social and psychological facts. Of course, raising these disregarded points also reveals a mental model. Then, the point should not be the existence or not of a mental model – it always exists – but its adequacy or inadequacy for a specific task.

In summary, events are analyzed from a dominant logic that works by filtering information. If the dominant logic is technical, most of the accumulated learning will appear as regulatory or technological developments.

The dominant logic will determine a mental model that is variable in different collectives and different moments. The same clues are not available when an event is analyzed, when it happened or when someone tries to prevent a future event.

The analysis of an event takes place under the model of thinking backward, while the generation of new regulations and systems happens under the model of thinking forward. Finally, the reaction during the event is produced in the thought modality of limited rationality and heuristic processing.

For a technical point of view, the most fragile modality is that used during the event. Then, the most "rational" modalities are oriented to the generation of rules with predesigned solutions for the event instead of leaving the management of such an event to people.

This common practice should not be criticized. Moreover, it is in the foundation of human knowledge development, as clearly identified by Thomas Sowell (1980):

> Civilization is an enormous device for economizing on knowledge. The time and effort (including costly mistakes) necessary to acquire knowledge are minimized through specialization, which is to say through drastic limitations on the amount of duplication of knowledge among the members of society.

However, if this practice is taken to its maximum level, some unintended effects could appear.

The same practice or device used to anticipate an event could prevent the management of other. It is true that operators do not have all the data that investigators have, but they have contextual information that a regulation or a technological design could have ignored.

Heuristic processing models – typically human – appear after trying to apply a general model. If this model does not provide a valid answer, the comparison of the current situation with individual experiences and relevant knowledge takes place. Then, the decisions will be aligned with the context.

Obviously, predesigned general solutions don't have this "emergency resource" at their disposal, while the human operator does. An example will demonstrate the point.

> US1549 (2009), known as the Hudson River landing, has been widely investigated, and some surprising facts have appeared: The manufacturer never foresaw a double engine stop below 20,000 ft. Hence, the checklist for full loss of power was far longer than practicable at low altitude.
>
> The operators, conscious of the problem, did not perform the checklist in a standard way and they jumped to some critical items such as starting the APU to keep the systems energized.

A "rational" approach in the example should not be able to manage the situation inside the constraints imposed by design. However, someone could still advocate for an extreme position such as, for instance, having an automatic system to perform the checklist in a time far shorter than that required by human pilots.

Again, that approach clashes with the contextual information: The checklist process would be extremely fast but based on sensors. If any of them fails (remember the QF32 case with its falsely failed engine), the input would not be questioned from observation nor from basic knowledge because none of them would be available for the computer.

Actually, cases such as the aforementioned AF447 or XL888T happened because the behavior of the people involved could be compared with that expected by a computer – that is, confusion in the face of an automatic reaction that was answering as designed to a bizarre input.

Predesigned solutions only work before foreseen contingencies and, by definition, in a complex environment, not all contingencies can be foreseen. Therefore, predesigned solutions should never be the only resource.

In summary, factual events, if captured, are a major improvement resource as good as foreseen potential events. However, the right lesson should be extracted from them and the approaches to some events suggest that the system works under a one-size-fits-all model. Therefore, it can work perfectly – as already shown – in some events, while it can have serious shortcomings in the management of others.

Notes

1. This mechanism is used during landing and consists of reversing the flow of the engine exhaust, using this to brake the airplane.
2. The QF32 case is an excellent reference for disagreements and open discussions in the cockpit. Of course, a single pilot would miss a major part of the process that led to a successful outcome.
3. There is still controversy about the real cause. It can be found here: https://youtu.be/DF68-HQ74tI.
4. Electrical wiring interconnection system.
5. Data provided by the captain of the airplane, D. José Luis Acebes de Villegas.

Bibliography

Air Accidents Investigation Branch (1973). Civil aircraft accident report 4/73 Trident I G-ARPI: Report of the public inquiry into the causes and circumstances of the accident near Staines on 18 June 1972.

Australian Transport Safety Bureau (2013). AO-2010-089 Final investigation: In-flight uncontained engine failure Airbus A380-842, VH-OQA overhead Batam Island, Indonesia, 4 November 2010.

Aircraft Accident Investigation Committee Thailand (1993). Lauda Air Luftfahrt Aktiengesellschaft Boeing 767-300ER Registration OE-LAV Dan Chang District Suphan Buri Province Thailand 26 May B.E. 2534 (A.D. 1991).

Aviation Safety Reporting System (ASRS) database (2019). ASRS Database Online. https://asrs.arc.nasa.gov/search/database.html.

Aviation Safety Network (2007). Turkish Airlines Flight 981. Accident report, accessed November, 23.

Beaty, D. (1991). *The Naked Pilot: The Human Factor in Aircraft Accidents*. Shrewsbury, England: Airlife.

Beck, U., & Rey, J. A. (2002). *La sociedad del riesgo global.* Madrid, Spain: Siglo Veintiuno.
Bureau d'Enquêtes et d'Analyses pour la sécurité de l'aviation civile (1993). Official report into the accident on 20 January 1992 near Mont Sainte-Odile (Bas-Rhin) of the Airbus A320 registered F-GGED operated by Air Inter.
Bureau d'Enquêtes et d'Analyses pour la sécurité de l'aviation civile (2002). Accident on 25 July 2000 at La Patte d'Oie in Gonesse (95) to the Concorde registered F-BTSC operated by Air France.
Bureau d'Enquêtes et d'Analyses pour la sécurité de l'aviation civile (2010). Report on the accident on 27 November 2008 off the coast of Canet-Plage (66) to the Airbus A320-232 registered D-AXLA operated by XL Airways Germany.
Bureau d'Enquêtes et d'Analyses pour la sécurité de l'aviation civile (2012). Final report on the accident on 1st June 2009 to the Airbus A330-203 registered F-GZCP operated by Air France flight AF 447 Rio de Janeiro–Paris.
Bureau d'Enquêtes et d'Analyses pour la sécurité de l'aviation civile (2016). Final report on the accident on 24 March 2015 at Prads-Haute-Bléone (Alpes-de-Haute-Provence, France) to the Airbus A320-211 registered D-AIPX operated by Germanwings.
C. A. O. (1951). ICAO Annex 13 (A number of additions relating to the status to be granted to investigation records and coordination between the investigator-in-charge and judicial authorities, were introduced in the mid-70s).
Comisión de Investigación de Accidentes e Incidentes de Aviación Civil (1977). A-102/1977 y A-103/1977. Accidente ocurrido el 27 de Marzo de 1977 a las aeronaves Boeing 747, matrícula PH–BUF de K. L. M. y aeronave Boeing 747, matrícula N736PA de PANAM en el Aeropuerto de los Rodeos, Tenerife (Islas Canarias).
Comisión de Investigación de Accidentes e Incidentes de Aviación Civil (1982). Technical report: Accident occurred on September 13th 1982 to McDonnell Douglas DC10-30-CF aircraft Reg. EC-DEG at Malaga Airport.
Comisión de Investigación de Accidentes e Incidentes de Aviación Civil (1986). Accidente ocurrido el 7 de diciembre de 1983 a las aeronaves McDonnell Douglas DC9 y Boeing B-727-200, matrículas EC-CGS y EC-CFJ, en el aeropuerto de Madrid-Barajas.
Comisión de Investigación de Accidentes e Incidentes de Aviación Civil (2011). A-032/2008 Accidente ocurrido a la aeronave McDonnell Douglas DC-9-82 (MD-82), matrícula EC-HFP, operada por la compañía Spanair, en el aeropuerto de Barajas el 20 de agosto de 2008.
Congressional Research Service (2003). NASA's Space Shuttle Columbia: Synopsis of the report of the Columbia Accident Investigation Board.
EASA, CS25 (2018). Certification specifications for large aeroplanes, amendment 22. https://www.easa.europa.eu/certification-specifications/cs-25-large-aeroplanes.
Einhorn, H. J., & Hogarth, R. M. (1981). Behavioral decision theory: Processes of judgement and choice. *Annual Review of Psychology*, 32(1), 53.
Gabinete de Prevençao e Investigaçao de Acidentes com Aeronaves (2001). Accident investigation final report all engines-out landing due to fuel exhaustion Air Transat Airbus A330-243 Marks C-GITS Lajes, Azores, Portugal, 24 August 2001.
Gigerenzer, G. (2007). *Gut Feelings: The Intelligence of the Unconscious.* New York: Penguin.
Gigerenzer, G., & Todd, P. M. (1999). *Simple Heuristics That Make Us Smart.* Oxford: Oxford University Press.

Hale, A. R., Wilpert, B., & Freitag, M. (Eds.) (1997). *After the Event: From Accident to Organisational Learning.* New York: Elsevier.
House of Commons (2009). The Nimrod Review: An independent review into the broader issues surrounding the loss of the RAF Nimrod MR2 Aircraft XV230 in Afghanistan in 2006 Charles Haddon-Cave QC report.
Khakzad, N., Khan, F., & Amyotte, P. (2012). Dynamic risk analysis using bow-tie approach. *Reliability Engineering and System Safety,* 104, 36.
Luhmann, N. (1993). *Risk: A Sociological Theory. A.* New York: de Gruyter.
Maturana, H. R., & Varela, F. J. (1987). *The Tree of Knowledge: The Biological Roots of Human Understanding.* New Science Library/Shambhala.
Maurino, D. E., Reason, J., Johnston, N., & Lee, R. B. (1995). *Beyond Aviation Human Factors: Safety in High Technology Systems.* Aldershot: Ashgate.
Maurino, D., & Canadian Aviation Safety Seminar (2005, April). Threat and error management (TEM). Canadian Aviation Safety Seminar, Vancouver, Canada. http://flightsafety.org/archives-and-resources/threat-and-error-management-tem.
Ministry of Transport and Civil Aviation (1955). Report of the Court of Inquiry into the accidents to Comet G-ALYP on 10th January 1954 and Comet G-ALYY on 8th April 1954.
National Aeronautical and Space Administration (1986). Report to the President by the Presidential Commission on the Space Shuttle *Challenger* accident.
National Transportation Safety Board (1990). NTSB/AAR-90/06 United Airlines Flight 232 McDonnell Douglas DC-I0-10 Sioux Gateway Airport Sioux City, Iowa, July 19, 1989.
National Transportation Safety Board (1996). NTSB/AAR-96/01 In-flight icing encounter and loss of control Simmons Airlines, d.b.a. American Eagle flight 4184 Avions De Transport Regional (ATR) model 72-212 Roselawn, Indiana, October 31, 1994.
National Transportation Safety Board (2000). NTSB/AAR-00/03 In-flight breakup over the Atlantic Ocean, Trans World Airlines flight 800 Boeing 747-131, N93119, near East Moriches, New York July 17, 1996.
National Transportation Safety Board (2002). NTSB/AAB-02/01 Aircraft accident brief EgyptAir Flight 990 Boeing 767-366ER, SU-GAP 60 miles south of Nantucket, Massachusetts, October 31, 1999.
National Transportation Safety Board (2010). NTSB/AAR-10/03 Loss of thrust in both engines after encountering a flock of birds and subsequent ditching on the Hudson River, US Airways Flight 1549 Airbus A320-214, N106US Weehawken, New Jersey, January 15, 2009.
National Transportation Safety Board (2019). Aviation Accident Database & Synopses. https://www.ntsb.gov/_layouts/ntsb.aviation/index.aspx.
Netherlands Aviation Safety Board (1992). Aircraft accident report 92-1 1 El Al Flight 1862 Boeing 747-258F 4X-AXG Bijlmermeer, Amsterdam, October 4, 1992.
Office of Air Accident Investigations NZ (1980). Report 79/139 Air New Zealand McDonnell Douglas DC10-30 ZK-NZP Ross Island Antarctica, 28 November 1979.
Perrow, C. (1972). *Complex Organizations: A Critical Essay.* New York: McGraw-Hill.
Petroski, H. (1992). *To Engineer Is Human: The Role of Failure in Successful Design.* New York: Vintage.
Popper, K. R. (1977). *Búsqueda sin término: Una autobiografía intelectual.* TecnosBarcelona.
Reason, J. (2016). *Managing the Risks of Organizational Accidents.* Hampshire, England: Ashgate.

Reason, J. (2017). *The Human Contribution: Unsafe Acts, Accidents and Heroic Recoveries*. Boca Raton, FL: CRC Press.
Risukhin, V. (2001). *Controlling Pilot Error: Automation*. New York: McGraw-Hill.
Rudolph, J., Hatakenaka, S., & Carroll, J. S. (2002). Organizational learning from experience in high-hazard industries: Problem investigation as off-line reflective practice. MIT Sloan School of Management Working Paper 4359-02.
Sánchez-Alarcos Ballesteros, J. (2007). *Improving Air Safety through Organizational Learning: Consequences of a Technology-Led Model*. Hampshire, England: Ashgate.
Shappell, S. A., & Wiegmann, D. A. (2000). The Human Factors Analysis and Classification System (HFACS). https://commons.erau.edu/publication/737.
Sowell, T. (1980). *Knowledge and Decisions* (Vol. 10). New York: Basic Books.
Stanton, N. A., Harris, D., Salmon, P. M., Demagalski, J., Marshall, A., Waldmann, T., and Young, M. S. (2010). Predicting design-induced error in the cockpit. *Journal of Aeronautics, Astronautics and Aviation*, 42(1), 1.
The Dutch Safety Board (2010). Number M2009LV0225_01: Crashed during approach, Boeing 737-800, near Amsterdam Schiphol Airport, 25 February 2009.
Transportation Safety Board Canada (1992). Commission of inquiry into the Air Ontario Crash at Dryden, Ontario: Final report. The Honourable Virgil P. Moshansky, commissioner.
Walters, J. M., Sumwalt, R. L., & Walters, J. (2000). *Aircraft Accident Analysis: Final Reports*. New York: McGraw-Hill
Weick, K. E., & Sutcliffe, K. M. (2011). *Managing the Unexpected: Resilient Performance in an Age of Uncertainty* (Vol. 8). San Francisco: John Wiley.
Wiegmann, D. A., & Shappell, S. A. (2003). *A Human Error Approach to Aviation Accident Analysis: The Human Factors Analysis and Classification System*. Surrey, England: Ashgate.

3

The Changing Roles of Technology and People in Aviation

The main issues in this chapter are the roles of people and technology, how they have changed over time, how we should expect them to change and what shortcomings could be behind that expected change.

As an example, Wells (op. cit.) mentions technology improvements addressed to navigation and traffic control precision, including landings with zero visibility, improvement in materials, engine reliability and information systems in the cockpits.

Very often, efficiency reaps the benefits of technology improvements originally addressed to safety. Some examples will show this effect.

1. An increase in navigation precision increases safety, but it can be used to decrease the distance between flying planes.
2. Better weather information can help planes to avoid entering storms, but it can be used to navigate them near to the cells without entering them.
3. Improvements in landing systems can be used to achieve safer landings, but it can also be used in airports that would otherwise remain closed.
4. Improvements in engine reliability can decrease the number of engine stops, but it can be an invitation to decrease the number of engines on a plane.
5. Better altimetry decreases the risk of midair collision, but it can be used to decrease the distance between planes.
6. Secondary radar decreases the collision risk through better information, but it can be used to put more planes into the same airspace.

A common path can be found in all these cases: Technology increases the safety level by reducing the chances of a negative event. However, things don't stop there. Instead, the safety level that has been added to the system is immediately invested in gaining efficiency.

Since this common trade-off causes safety levels to decrease again, it seems clear that the real improvement engine is not safety but efficiency.

Human Error: Myths and Reality

Technology improves while the human role has long been questioned. That's an old story, and new developments in Artificial Intelligence (AI) and unsupervised Machine Learning do not decrease that trend, rather the opposite.

Boeing pointed out in its "Statistical Summary of Commercial Jet Airplane Accidents" (2006) that 55% of accidents were produced by the behavior of the flight crew (Figure 3.1).

More recent editions have changed the classification, since the old one – despite being present in many activities – hides some facts that could bias the role of people in the system.

A single accident can have concurrent causes. The analysis could then be misguided if centered on human error (Mauriño et al., op. cit.). Additionally, there are two reasons to treat the old Boeing publication with a certain amount of skepticism, even when accepting the data.

- If 55% of accidents come from flight crew errors, that could give a false picture: If we find a way to suppress people in the system, accidents would decrease within that 55%.
- Human error usually appears together with other problems that have a major role in the outcome.

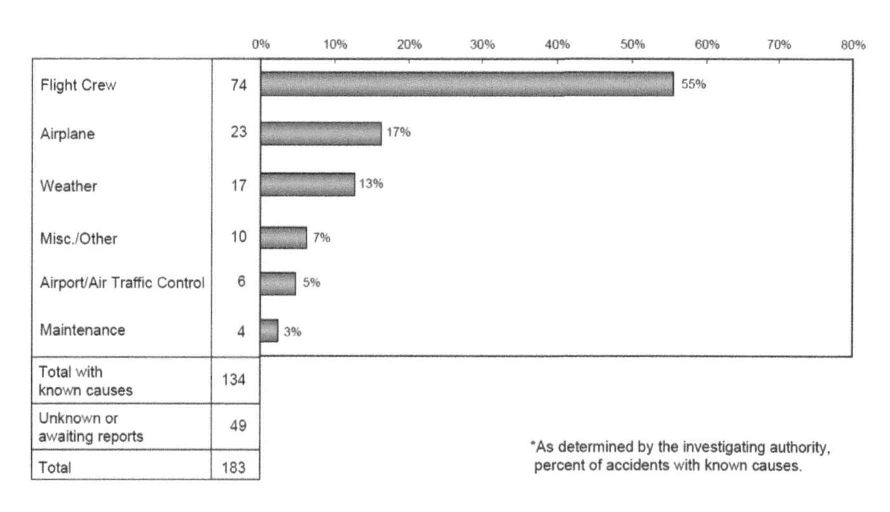

FIGURE 3.1
Accidents by primary cause. (From © 2006 The Boeing Company. All Rights Reserved.)

For instance, the inadequate behavior of a pilot after an engine stop can lead to an accident. However, even if the error is beyond any doubt, it seems equally clear that a technical fault (i.e. the engine stop), started the sequence leading to the accident.

As Leveson (op. cit.) points out:

> The same event can give rise to different types of links according to the mental representation the analyst has of the production of this event. When several types of rules are possible, the analyst will apply those that agree with his or her mental model of the situation.

Then, from a technical point of view, someone could say that the engine stop also came from human error, because of poor maintenance or design. If so, that could be so true as to be useless.

The resulting statement – that 100% of accidents come from human error – does not add to the analysis anything but noise. Any chance for a causal analysis disappears.

The social impact of an accident impedes the visibility of the other extreme of the continuum – the most relevant one – represented by accidents avoided by the sound performance of people.

Furthermore, if we speak about how people avoid accidents, we should introduce a differentiation:

1. Situations where people behave as expected after an event – that is, people behaving as a part of the system as designed: In these cases, even if the accident is avoided through the human element, the credit for the avoidance should go to the system, not the human.
2. Situations where people must manage unforeseen events: Since there is no guide or procedure for an unforeseen situation, the full credit for avoiding an accident should go to the human element.

Evaluating this second option is difficult, since actions taken to avoid an accident are usually not as dramatic as QF32, US1549, AA96, AC143, U232 and many other cases.

Very often, the action seems trivial and is taken in an early phase; it can imply managing known bugs in the system and breaks the chain of events; hence, it can go unreported.

In summary, many unforeseen events are not emergencies at all, and they never reach that level because of human intervention. The importance of this factor is greater than expected. Usually, these events that are "below the radar" pass unreported. Some systems already in place – those devoted to line training or to the detection of threats – could be used to detect these low-intensity/high-potential events, but these systems are typically not aimed at this kind of detection.

Overcoming system failures is "business as usual" for pilots. They don't report small failures because they are familiar with them and reporting them would be harder than managing them. However, many of these small unreported events could lead to major events if not properly attended to:

1. A faulty radio altimeter linked to an automatic system produced an accident (TK1951). However, the same fault had appeared in different flights and the sequence was stopped by switching to manual landing.
2. A programmed approach to a major airport could be erased from the flight management system in conditions incorrectly read by the plane systems as overflying the airport instead of approaching it.
3. Lack of communication in oceanic flights is considered a temporary condition, not an emergency.
4. Different systems can give false warnings that, due to their frequency, are ignored when the conditions lead the crew to think that the situation does not exist.
5. Bizarre behaviors by navigation systems that are usually only reported when they lead to a near-miss.

Many other common situations could be added, but since that information is unavailable or unused, decisions can be made using incorrect inputs or leading to false conclusions – that is, conclusions coming from the common myth about how the system would improve by removing the human element.

Major errors and major achievements are usually recognized, and people close to the event become heroes or villains, but actions that are apparently minor but which help to avoid a major event commonly go undetected. Hence, analyses about the convenience of having a single pilot or, directly, self-driven planes don't consider one point:

The external environment is complex enough to surprise pilots with new and unexpected situations. However, the internal environment has become complex enough to be a source for different unexpected situations.

Adding more complexity to the already complex algorithms used in different systems is like using a thicker rope to fix another rope whose own weight was causing it to break.

The Human Role: Skills and Knowledge as Accident Triggers

People can have two different but related roles in a system:

In the first role, people are told what to do in many different situations through specific procedures. This is the role that is being progressively taken by technology.

In the past, many planes required two pilots, a flight engineer, a radio operator and a navigating officer. Now, there are few exceptions to the general rule of having only two pilots.

The navigating officer disappeared long ago, once land-based resources were sufficient to support the pilot. The next one to disappear was the radio operator, once technology became more reliable and radios easier to handle. Finally, the flight engineer is absent from practically every fleet, as the tasks linked to that position are by automated.

Furthermore, there is ongoing research, endorsed by the U.S. Senate, to assess the feasibility of having a single pilot on cargo planes or even big cargo drones flying only-cargo routes:

> The FAA, in consultation with NASA, shall conduct a review of FAA research and development activities in support of single-piloted cargo aircraft assisted with remote piloting and computer piloting.

Many of the tasks that justified the presence of the former crowd on a plane have simply disappeared, through automation or through radical changes.

The cockpit of the first B747 had about 900 elements (controls and indicators), which decreased to about 300 in the 747-400 generation – a cleaner environment. That difference is due to automation but also to the way the elements are counted: The existence of multifunctional screens – which can be counted as a single element – with long streams of options, some of them hard to access, leads to false accounting.

Anyway, this is a well-known process defining the evolution shared by Aviation and many other fields. However, we should ask whether this evolutionary path could have some side-effects and what next steps should be expected.

Again, we will find efficiency to be the driver of the entire learning process. Automating tasks is a way to achieve high performance with less people and less training. Actually, this is very often the main payback for the investment in technological resources.

Some effects of this process are already widely visible at the present stage. Long ago (2004), CAA-UK issued a paper called "Flight Crew Reliance on Automation", with some warnings about this issue.

> Modern large transport aircraft have an increasing amount of automation and crews are placing greater reliance on this automation. Consequently, there is a risk that flight crew no longer have the necessary skills to react appropriately to either failures in automation, programming errors or a loss of situational awareness. Dependence on automatics could lead to crews accepting what the aircraft was doing without proper monitoring. Crews of highly automated aircraft might lose their manual flying skills, and there is a risk of crews responding inappropriately to failures.

We could say that these long-identified issues are still alive and thriving in the present. Furthermore, this issue is behind something that could be called *new-generation accidents*, with cases like XL888T, AF447, Asiana 214 and the very recent Lion Air 610.

Additionally, many professional pilots in long-haul fleets perform three or four landings a month, not including those performed in simulators for training purposes. So, as the aforementioned cases and others have shown, many pilots don't feel comfortable flying the plane manually, as this extract from the official report shows.

> A NOTAM had been published indicating that the ILS glideslope for runway 28 L was out of service due to a construction project, and the flight crew was aware of the outage. While the electronic vertical guidance provided by a glideslope is required for approaches in certain low visibility conditions and can be a useful aid in all weather conditions, a glideslope is not required for a visual approach. The flight crew had numerous other cues to assist in planning and flying an appropriate vertical flightpath to the runway. Flight crews routinely plan descents based on speed and distance from airports or navigational fixes and published crossing altitudes on approach charts. Aids such as the ND's green altitude range arc and VNAV features were available to guide the pilots in the initial portion of the descent. As the airplane neared the runway, the PAPI lights and the visual aspect of the runway surface provided additional cues. The NTSB concludes that although the ILS glideslope was out of service, the lack of a glideslope should not have precluded the pilots' successful completion of a visual approach.

This lack of manual practice was more recently identified in a report by FAA ("Operational Use of Flight Path Management Systems", 2013) but, in this case, the concern was beyond manual flight (Figure 3.2).

The issue is not only manual flight skills but *insufficient knowledge*. Both come in the same package and have the same origin – that is, inadequate design and use of automation.

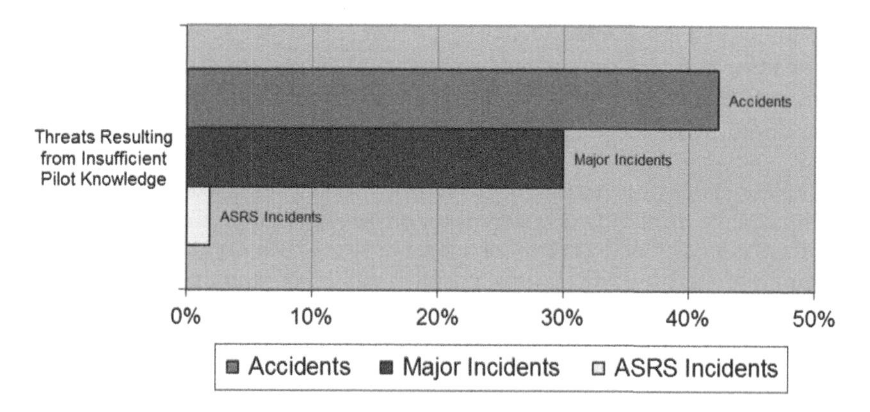

FIGURE 3.2
Threats resulting from insufficient pilot knowledge. (From PARC/CAST Flight Deck Automation WG, Final report: Operational use of flight path management systems, 2013.)

One of the main manufacturers, Airbus, has already expressed its concerns about the problem and proposed changes in the training process. After the presentation of its A350 model, these changes were communicated in a *Wall Street Journal* interview:

> The goal is to first "just have them feel the plane, and how it behaves without" turning on automation or presenting any complicated system failures or emergencies....
>
> Eventually, Airbus seeks to expand the revamped training approach to other models....
>
> The new focus is the strongest sign yet of industry-wide concern about the hazards of excessive reliance on automated cockpits and worries about pilots who may be reluctant to take over manual control when necessary. The result could be to accelerate the movement of airlines toward training programs highlighting manual flight maneuvers.

Therefore, there is a convergence between both the uses of human operators: They have tasks related to normal operations and they are an alternative resource for abnormal situations. However, the automation of normal operations decreases the opportunities to maintain current the skills required for abnormal situations.

Perhaps the first question should be about the right level of automation. There seems to be a default answer to that question: as much as technology can provide.

Different events show that this apparently default answer could be wrong. Maintaining situation awareness and, hence, the full capacity to react to unexpected events requires having the pilot in the loop, not as a spectator but involved in the control.

Full automation is technically feasible but perhaps it is a mistaken goal if, at the same time, the capacity to manage incoming events is reduced.

Right now, updating manual skills is the role of simulator training instead of being a by-product of normal operations. Furthermore, the proposed change, aimed at recovering manual skills, will require changes even in the training process, very often more focused on systems management than on flying the plane.

Nowadays, state-of-the-art simulators have reached the point where their imitation of a real plane is almost perfect. Hence, we should not have any objections to keeping piloting skills current in that way. However, some issues remain:

1. Airlines prefer to have their pilots expending time on real flights than in training sessions.
2. Many airlines want their pilots to use automatic processes as much as possible.

3. Training sessions are usually foreseeable by the pilot. There is an absence of the "startle effect" that can happen in real flight. This factor has been widely criticized because it puts a question mark over the effectiveness of the training process.

> The issue with foreseeable situations became public after the movie *Sully*. The information below is taken from the official report, showing the importance of previous knowledge, saving the time for situation assessment:
>
> The simulations were conducted using an Airbus A320 full-motion, pilot-training simulator and a fixed-base engineering simulator to determine whether the accident airplane could have glided to and landed at LGA or TEB after the bird strike, considering both an immediate return to LGA and a return after a 35-second delay ... The simulators were programmed to duplicate as closely as possible the conditions of the accident flight, including winds, temperature, altimeter setting, and weight and balance. The profile flown duplicated as closely as possible the accident profile ... The pilots were fully briefed on the maneuver before they attempted to perform it in the simulator. The following three flight scenarios were flown: (1) normal landings on runway 4 at LGA, starting from an altitude of 1,000 or 1,500 feet on approach; (2) attempted landings at LGA or TEB after the bird strike, starting both from zero groundspeed on takeoff from runway 4 at LGA and from a preprogrammed point shortly before the bird strike and loss of engine thrust; and (3) ditching on the Hudson River starting from 1,500 feet above the river at an airspeed of 200 kts. During the first flight scenario, all of the pilots were able to achieve a successful landing in both simulators ... Regarding the second flight scenario, 20 runs were performed in the engineering simulator from a preprogrammed point shortly before the loss of engine thrust in which pilots attempted to return to either runway 13 or 22 at LGA or runway 19 at TEB ... In eight of the 15 runs (53 percent), the pilot successfully landed *after making an immediate turn to an airport after the loss of engine thrust.*

Some recent accidents show a mix of lack of manual flying skills together with a poor understanding of the systems, how they react and what should be expected – that is, a low situation awareness.

Cases like TK1951 and the aforementioned Asiana 214 show that mix. Both crashed on the final approach and, although the detailed reasons were different, both cases had something in common:

The pilot did not know exactly what the plane was doing and why. The Asiana case had something particularly related to manual skills: A failure in the ground equipment of the instrumental landing system required the pilot to perform a manual landing, and this was disturbing for him.

Turkish Airlines 1951 was intended to land in Amsterdam using the automatic procedure. A faulty radio altimeter gave a false input to the autopilot, which set the power to idle. Low visibility together with unchecked automation led to a crash near to the runway.

Since this same failure happened on other planes that avoided a bad outcome, it seems that the event could have been avoided through alertness, with better visibility or both. The official report includes two paragraphs very clearly pointing to the lack of system understanding by the pilots:

> The Board concludes that the improper functioning of the left-hand radio altimeter system led to the thrust from both engines being reduced by the autothrottle to a minimal value too soon, ultimately causing too big a reduction in speed. The airspeed reached stall speed due to a failure of monitoring the airspeed and pitch attitude of the aircraft and a failure to implement the approach to stall recovery procedure correctly. This resulted in a situation where the wings were no longer providing sufficient lift, and the aircraft crashed.
>
> ...The erroneous altitude reading ... was used by various aircraft systems, including the autothrottle. The crew were unaware of this, and could not have known about it. The manuals for use during the flight did not contain any procedures for errors in the radio altimeter system. In addition, *the training that the pilots had undergone did not include any detailed system information that would have allowed them to understand the significance of the problem.*

In Asiana 214, together with the manual skills problem, the official report pointed out the lack of knowledge about how the automation worked.

> Reduced design complexity and enhanced training on the airplane's autoflight system. The PF had an inaccurate understanding of how the Boeing 777 A/P and A/T systems interact to control airspeed in FLCH SPD mode, what happens when the A/T is overridden and the throttles transition to HOLD in a FLCH SPD descent, and how the A/T automatic engagement feature operates. *The PF's faulty mental model of the airplane's automation logic led to his inadvertent deactivation of automatic airspeed control.* Both reduced design complexity and improved systems training can help reduce the type of error that the PF made.

Of course, the misunderstanding visible in these cases and made explicit by the official reports appears in many others, some of them already mentioned, such as AF447 and XL888T. The misunderstanding also goes beyond the cockpit to include the maintenance area. A clear and extreme case of misunderstanding in maintenance is the G-KMAM case. Again, the official report explains clearly what happened but, in this case, a new element is introduced:

> The incident occurred when, during its first flight after a flap change, the aircraft exhibited an undemanded roll to the right on takeoff, a condition which persisted until the aircraft landed back at London Gatwick Airport 37 minutes later. Control of the aircraft required significant left sidestick at all times and the flight control system was degraded by the loss of spoiler control.

Another paragraph reports the lack of knowledge by the pilots:

> The operator had not specified to its pilots an appropriate procedure for checking the flight controls.

Both paragraphs are descriptive of the event but, by far, the most illustrative about a general situation, not only for the G-KMAM case but in general, is the next one:

> While the potential for error has always existed, even with simple aircraft types, the skill and experience of those faced with day to day problems may well have been sufficient to achieve a safe conclusion. With the introduction of aircraft like the A320, A330, A340 and Boeing 777, *it is no longer possible for maintenance staff to have enough information about the aircraft and its systems to understand adequately the consequences of any deviation*. The avoidance of future unnecessary accidents with high technology aircraft depends on the attitude of total compliance within the industry being developed and fostered. Maintenance staff cannot know the consequences of non-standard operations on the system.

The intention of emphasizing procedure compliance is clear in this paragraph. However, in doing so, it introduces a kind of collateral damage.

At the same time, the report gives a good reason to avoid any deviation *and* it introduces as an unquestionable matter of fact the idea that *the knowledge of maintenance staff is limited to the skills required to follow the procedures*. They cannot be questioned by those who must perform them because they don't adequately understand.

Of course, this situation, perfectly explained in an official report about a maintenance error, is shared by pilots, and the lack of functional knowledge related to the interaction of the systems is seen as normal.

This issue is far from being new; it can be found even in a weird accident that occurred in 1973 (National Airlines 27), where an uncontained explosion took place in one of the engines, as explained by the official report:

> Just before the explosion, [the captain] and the flight engineer had discussed the electronic interrelationship between the autothrottle system and the associated N1 tachometers. As a result of their discussion, it was decided to check certain functions of the system. The captain stated that 'The flight engineer and I were speculating about where the automatic throttle system gets its various inputs, whether it came from, for example, the tachometer itself, the N1 tachometer, or from the tachometer generator. So we set up the aircraft in the autopilot and in the airspeed (autothrottle) mode … allowed the airspeed to stabilize (at the preselected 257 KIAS) then selectively, successively pulled the N1 circuit breakers on 1, 2, 3 engine.' He further stated, 'We retained a speed mode in the enunciator. I was satisfied at that point that the pick up came at some other point than the gage itself, but to check further, I retarded the speed bug on the airspeed indicator slightly … I merely wanted to see if

> the throttle followed the speed bug. I backed up the speed bug approximately 5 knots, and noticed that the throttles were retarding slightly. I reached in and disengaged the autothrottles and turned to the engineer and made some remark to him that I was satisfied with this function and at that point the explosion took place.

The fact that the captain and the flight engineer were guessing about the data input for one of the most commonly used systems in the plane shows something: They did not know.

A lack of information, or incorrect information, about the interaction among different systems is a feature shared by many different accidents. So, the human role, including required skills and knowledge, should be reviewed. Otherwise, we can enter – or remain – into a negative dynamic:

1. A bigger and bigger part of the knowledge required to handle a plane is embodied in technology and procedures. People are reduced to performing those tasks that are difficult or expensive to automate.
2. When an emergency appears, people lack an understanding of the system, both general and situation specific, and the skills to manage it.
3. So, people conclude that the human contribution is negligible, and the next step should be removing one of the pilots from the cockpit.

Therefore, the evolution of the system creates the conditions for people to become unfit to handle a serious event. Once there, the designers suggest removing people from the system because they are unfit. Perhaps a different solution should be found instead.

The Human Role: Skills and Knowledge in Accident Avoidance

Lack of knowledge and skills is behind some accidents. Actually, this lack of knowledge and skills appears as a loss of situation awareness, whether it comes from a lack of basic knowledge or from confusion in a complex and unforeseen situation.

Confusion is present in many accidents affecting technologically advanced planes; that is, people involved in the event did not know what was going on.

At the same time, there is an opposite side, composed of events that did not become major accidents because of the actions of people. In these cases, an uncommon level of situational awareness can be found.

The aforementioned QF32 and US1549 cases show, above all, an exceptional level of situational awareness. More disasters-to-be that could qualify in the same category are, for instance, AA96 and U232.

They shared a common feature: The situation was unthinkable since the redundancy of the systems supposedly guaranteed that the plane would always be safe. Actually, DC10 (U232 and AA96) had three hydraulic systems working perfectly when the plane took off, while ETOPS rules authorize transoceanic flights with two engines. In the same fashion, an A380 (QF32) would show that even a four-engine plane can suffer a very serious event stemming from a major fault in only one of them, and an A320 (US1549) would show that a full loss of power can happen just after take-off.

All the aforementioned events were managed with solutions coming from basic knowledge, and all of them happened in situations where operating procedures became useless.

On American Airlines 96 and United 232, the *eureka* moment came when the pilots, on a plane uncontrollable by conventional means, discovered that they could regain some control playing with the power levers.

To get there, they used a piece of basic knowledge: Full power can make the nose to go up, while decreasing power can make the nose to go down. Adding more power to one of the engines can make the wing of that side go up, starting a turn.

In US1549, the pilots became conscious that, once the systems lose energy, they could lose many of the systems. Using the auxiliary power unit – a little turbine used to start the engines – as a generator, they could keep the systems alive.

Furthermore, calculations to know if they could reach the airport or not were complex, especially on the go. However, there is a common pilot heuristic to use instead of that calculation: Getting a fix point in the windshield. If the fix point goes up the windshield, the plane will not reach that point; if the fix point goes down, the plane will overfly the point. That heuristic is very common for pilots who land in unprepared places, but for airline pilots this is a piece of knowledge so curious as to be irrelevant to their daily practice.

QF32 is another situation of basic knowledge. Double unrelated engine stops are so infrequent that some twin planes are certified to fly more than 6 hours with one engine once the other has failed. Then, an independent failure of a second engine immediately after the first one would be very suspicious, especially if the second engine is in the opposite side of the plane across the fuselage.

Additionally, pilots could feel in the plane's behavior whether the engine was providing power or not. The conclusion: If the system was giving out false information, why should they suppose that this was the only false information? A general distrust in the data coming from the plane systems is established. From that moment, they would set aside the management of hundreds of warning messages and, instead, they would behave as if they were flying a normal plane.

BA38 showed another similar example: A double engine stop during the approach to Heathrow made it clear to the pilots that the plane would not reach the runway and they would crash. However, the final approach is usually flown with flaps down: Flaps allow the plane to lose altitude without gaining speed.

When the captain became conscious that they could crash against a building in the approach pathway and, since they had enough speed for a different setting, he raised the flaps. That would mean that the plane would go faster but it would lose altitude at a slower pace.

The solution was enough to avoid crashing against a building and, instead, they made a crash-landing very near to the runway. The case finished without fatalities.

Finally, the event known as the Gimli Glider: Due to an incorrect calculation, a Boeing 767 ran out of fuel and both engines stopped in flight. The only airport they could reach was a glider airfield with a short runway. Since their altitude was greater than required to land in the chosen place and they only had one opportunity to land, they performed a *sideslip*, common in gliders and biplanes but never seen before in a plane that size.

So, the cases with a happy outcome would show something especially important about the value of the human factor:

The specific knowledge used to regain control of the situation was not specific to the plane or the situation. Far from it, that knowledge could qualify as commonsense or background knowledge and, in the critical moment, it emerged as the key for the *eureka* moment that would set up the problem-solving process.

This fact will open new questions about the human contribution:

1. How should training be designed to guarantee that these pieces of basic knowledge, able to elicit the right answer in a critical situation, are present?
2. If we design an information system that is able to process a complex situation like these and to reach a similar solution, would that system be fast enough to solve the problem in the available time?

There is no clear answer to these questions. Furthermore, the research on the accidents focuses on the facts and the formal knowledge that people involved in the accident had. It is uncommon to have detailed information of them to devise where the knowledge came from or why it was absent.

There are few exceptions to that general rule and, among them, it was known that one of the pilots of the "Gimli Glider" was a glider pilot. Moreover, he was familiar with the airfield where they landed, and a maneuver as uncommon in Commercial Aviation as a sideslip would be present in his mind as active knowledge instead of a booklover's curiosity.

Right now, some pilots are trained *ab initio* to fly big planes, starting their training with simulators. When they start to transport passengers, they have never flown light airplanes and they have never flown alone. We could hardly expect them to find a solution like that the Gimli Glider case, since it would be fully out of their minds. Even so, they could know about the existence of such a maneuver, but they would never have performed it.

Another case where there is available information – beyond type ratings and so on – about the people behind a successful outcome is QF32. The captain provides some information about himself in the book named after the flight, *QF32*, and some interesting data appear.

From early youth, he was curious about how things worked, and he enjoyed assembling and disassembling devices as motorbike engines. When computers became ubiquitous, he started to play with them with the same curiosity. In the Aviation field, he had been a military pilot, experienced with different aircraft, including helicopters. This background could explain why he concluded that the information coming from the plane's systems was unreliable and why he decided that the best solution was going back to basics.

For many people trained in the currently accepted way, going back to basics would be the equivalent of going nowhere. This kind of information, considered to be a flaw of the reporting systems, is also missing in the reports, which are highly standardized about which data they must include and what kind of things are disregarded.

An open question such as "How did you figure out that solution?"[1] seems unthinkable, but it would be useful for refining the recruiting and training processes. The information should go well beyond licenses, ratings, dates and flown hours in both successful and unsuccessful outcomes.

The Human Contribution

Until now, we have focused on cases that show in an overly dramatic way at what point humans can be successful or not when performing specifically human activities.

We have not considered lapses and other mistakes that, being common, appear easier to manage by keeping some basic rules. For instance, confusing flaps and landing gear levers was a frequent event until someone observed that both levers were contiguous and that their shapes were similar.

Checklists have been used for a long time to prevent lapses, and procedures are supposed to encapsulate the best available practice to avoid dealing with known problems as new ones.

So, we should not consider certain tasks to be human contributions. These are tasks that, even if they are performed by humans, do not require from them any problem-solving process or any special skill but only to perform a task in a clear and prescribed way, and in the conditions that this task is supposed to be performed.

In some ways, these tasks are performed mechanically or by following a rule but do not imply a specifically human process unless something fails. However, before removing these low-value and mechanical tasks once technological development allows it, we should consider their contribution to keeping the human in the loop. Perhaps these tasks could have greater value than intended.

A classic experiment performed by Held and Hein in 1963 showed two cats sharing the same visual experience but a different activity: While one of them was moving, turning around a wheel, the other was in a cart towed by the first one. The working cat developed a normal visual perception while the second one, despite having the same visual inputs, developed a faulty perception.

Any human factors practitioner knows that people are better performers than supervisors. Hence, they would not accept statements like those that appeared after the first accidents involving cars with automatic pilots, which can be summarized in two sentences:

1. The automatic program drives better than you.
2. Do nothing except keep your hands on the wheel and stay alert.

That, simply, will not happen: Boredom and alertness are not compatible. So, defining who should oversee some clear and elementary tasks and whether or not they can be automated is not only a matter of technological feasibility:

Some tasks will not add value by themselves, but they will help to keep the pilot in the loop and, hence, keep the situation awareness at the right level. Those "low-value" activities will not define where the human contribution is, but eventually they could become pre-conditions to making that contribution feasible when required.

In the same way that transporting oil is not an objective for a car but a condition for its engine to run, some of these lower and easy-to-automate tasks should be kept instead of automated.

As an example, central advisory and warning systems (CAWS) are designed to prevent multiple warnings appearing when they are related to a single main issue. CAWS will inhibit the secondary warnings to avoid unnecessary and confusing noise. In general terms, the contribution will be positive and, if adequately designed, the system works fine in almost 100% of situations.

However, in some situations, the convergence of different issues could obscure the most critical of them. The next paragraph comes from the official report of Air Transat 236, where a plane landed with both engines stopped due to fuel exhaustion.

> At 5:33, a pulsing white ECAM advisory ADV message was generated and displayed in the memo area of the E/WD, indicating a 3,000 kg fuel imbalance between the right and left wing tanks. Under normal conditions, this ECAM advisory brings up the FUEL system page on the SD. However, *the manual selection of ENGINE systems page by the crew inhibited the display of the fuel page. A 3,000 kg fuel imbalance is an abnormal condition that does not result in a display of the corrective action required to correct the imbalance.* To ascertain the required corrective action, the crew must view the fuel page, diagnose the pulsing fuel quantity indications, and then refer to the appropriate page in the Quick Reference Handbook (QRH).

Beyond the fact that a system behavior that was supposed to add clarity could delay action in a critical event, a less automated process could provide early warnings about what is going on.

Being present in an early phase will provide missing cues to solve the problem before a full-blown emergency appears. The evolution of the event provides information, and before reaching the emergency phase, the pilot could receive useful data to solve the situation. Otherwise, a fully developed situation might appear out of the blue, making it harder to know what is going on. The following text, taken from the AF447 official report, emphasizes that point.

> When the Captain returned to the cockpit, the aeroplane was in a rapid descent, though at an altitude close to the cruise level it was at when he had left. *Under these conditions, and not having experienced the complete sequence of events, it was very difficult for the Captain to make a diagnosis.* He would have needed to question the co-pilots about the sequence of events, an approach that was blocked by the urgency of the situation and the stress conveyed by the PNF's tone of voice.

Then, perhaps one of the big issues in the present environment is that people are used like "information processors", disregarding the specific features that give them their real value, especially in high-risk environments.

Sociologist Edgar Morin said that the machine-machine was always superior to the man-machine. So, if designers insist in dealing with people as pieces of a machine, they will always find machines performing a better job.

Hence, if the system evolves by downgrading its human side, by decreasing recruiting and training requirements or by introducing systems whose operations are opaque to the user, or all of these things, nobody should be surprised that the machine performs better than the human.

Therefore, instead of creating situations that lead to low human performance and, hence, justifying the replacement of people with machines, the key should be defining the situations where specifically human features will be needed. The next step should be defining the process to have these features available when required.

For many designers, people are emergency resources to use when something fails. In the meantime, they can perform tasks that are difficult or expensive to automate; that is, they have a temporary assignment until technology finds a cheaper way for the task to be performed.

However, by doing that, people are kept out of the loop and, when emergencies arise, they are asked to awake and solve the situation. As we might expect, people can fail in that human-adverse environment.

There is the matter of training, of course, but not only training. Tasks and work environments must be defined in a way that will generate situation awareness and maintain alertness.

Regarding alertness, some designers have tried to find a workaround for the problem: introducing useless activities to keep people active. However, these activities, especially once identified as useless, don't meet the intended objective. A checklist or a position report from time to time is not the right way. At the same time, trends like the *dark cockpit* (displaying data only when they are outside the parameters) don't help to maintain situation awareness.

The main thing that justifies keeping people involved is the fact that they can question the environment and produce creative solutions. However, that only happens if the environment is designed to maintain full and continuous situation awareness in normal and abnormal situations.

People are expected to bring their "insight" to manage unforeseen situations, and some cases have been shown where that has happened. However, the emergence of the insight requires some conditions, from the system and from the people managing it.

The system must provide more than just a pleasing interface; it must provide functional information, where the user can go up and down an abstraction scale to define cause–effect links. Once these links are identified, the system must allow the user to act, instead of being constrained or misguided by automatic systems triggered by a sensor, because of a relationship that could be fully ignored by the average user with the average training.

Regarding the user side, if the system must provide functional information, the user must be able to understand that information well beyond "Windows knowledge". As Bennett and Flach established:

> Designing a representation to conform to the information processing limits of an operator without considering the consequences in terms of lost information with respect to the domain being controlled can end up trivializing the problem and can lead to a very brittle (unstable) control system.

That means an important change in a practice aimed at building complex systems with plain interfaces and, of course, that also means an important effort in recruiting and training. Gary Klein asked himself precisely why some people, sharing all the information with others, come up with the solution while others don't. The answer is closely related to training and to the features of the people who bring the solution, and, before going further, it can be anticipated that the suggested profiles will be quite different from those commonly used in recruiting.

From the information processing point of view, it is commonly accepted that computers are faster than people, but that is not always true:

People have processing models that can beat a computer because one of the features of human processing is a *stop rule* – that is, a rule to define when no more analysis is required. Additionally, people can jump to the right conclusions with far less data than that required by an I.T.-based processor, by recoding information in wider units and by eliminating unviable options instead of devoting time to processing them.

Not all places require human features to work, but high-risk activities performed in open environments do. Then, the environment should be designed to guarantee that these features are present and usable.

As already mentioned, Aviation has two specific properties that invite us to maintain the human contribution.

1. A situation cannot be frozen: Unlike other fields, where activity can be instantly stopped, the plane will keep flying at high speed and any potential event must be solved while flying.
2. Limited resources and people: Despite improvements in communication and external help from ATC, the crewmembers of a flying plane are basically alone. They can have higher-ranking officers, or some other people who might know better than them, but these people are not available in the situation. Decisions and subsequent actions are mainly local and have very limited support.

Nobody can foresee what the future will be and what might be the right mix of humanity and technology in different fields. Anyone trying to find the right approach should be aware of two issues:

1. Technology development: Many people are willing to be blinded by the brightness of AI, Machine Learning and so on, but things are not as advanced as they might pretend. However, there are remarkable achievements, and anticipating their real potential, once we have discounted the marketing noise, is hard.
2. People development: As explained in the foreword, this book is not against technology but against poorly designed technology that is not only unable to use the potential of people but leads them into situations where their worst side is expected to appear.

On the technology side, previous attempts at developing AI have committed a serious mistake: The AI pioneers underestimated the value of human intelligence. Supposedly, the human brain is a collection of evolutionary remains, and they expected better results from using a computer while ignoring how the brain works.

So, early adopters of AI tried to start from scratch to avoid the non-functional features that came from a slow and unplanned evolution.

Facts would show that this was not the right path to follow. Time and again, the human brain has beaten the fastest computers in some tasks, especially (but not only) poorly defined tasks in open environments.

Nowadays, the situation is different: A new generation of technologists is trying to analyze how the brain works and determine the features that lead to its superior performance in new or poorly defined situations. Once there, they should confront three challenges:

1. Processing: Developers should preserve those features in which computers are already superior to people while finding new ways to deal with situations where computer-like models are unfeasible (not computable) or very slow.
2. Learning: Unlike humans, computers can process millions of specific situations at an amazingly fast speed. The computer should imitate the human practice of looking for similar situations instead of general abstract models.

   However, processing that information in an organized way, so as to be sure that the relevant models are available, is difficult. Finding and storing information is easy; organizing it in the right way is harder.
3. Switching: Kahnemann refers to two systems: one of them specifically human and which frequently leads to error and the other one computer-like. Kahnemann warns us not to discard the first one; that is why we make mistakes when we are processing information in a computer-like way.

   We could find a similar problem in eventual AI developments: Perhaps computers will not be able to disconnect the sequential way in which they process information in situations that require exactly that.

These are not easy challenges for the new AI wave. Unlike what happened in the past, with so-called good old-fashioned artificial intelligence (GOFAI), it seems that this new wave is addressing the problem more directly. However, the problem that they confront – putting the positive side of human features into a computer – is amazingly complex and, for now, unfeasible.

Perhaps in the future, an intelligent system could be designed that is the best of both worlds, human-like and computer-like. Today, that is not feasible and human capabilities are required when new and unforeseen events – or, simply, those that are too complex for a computer-like approach – appear.

It is not possible to put humans into a emergency glass box and break them out when the situation requires it. That simply does not work. When designs are oriented that way, accidents have already shown that people miss essential inputs. Hence, people used as emergency resources do not work as planned in a poor design. So, people will fail, but the failure will be a product of a design that disregards human features.

Going back to one of the statements made after a fatal Tesla accident, published in *Business Insider*:

> So if you own or lease a Tesla, please do what I'm telling you: If you engage Autopilot, do not under any circumstances take your hands off the steering wheel. Treat the system like advanced cruise control and nothing more.

Many interface designers in different fields would agree with that statement, but regretfully humans do not work that way. It seems that, in the same way that there are user manuals for different systems, a user manual for people management should also exist. If so, that manual would say in clear terms that people cannot be asked to be bored but alert, supervising a system that displays selected items while keeping full situation awareness.

There seem to be implicit requirements in some environments to use people as emergency resources; as we have seen, that does not work. Maintaining the human contribution means keeping people active. Keeping people active means something more than giving them routine tasks that are easily performed by a computer, such as position reporting or routine checks. Otherwise, we can enter into a perverse but common logic about what the human contribution is:

1. We do not meet the basic conditions to maintain situation awareness.
2. Since situation awareness is lost, when an emergency appears, the management of the situation is poor.
3. Since the management of the situation is poor, we discover that the contributions of people are marginal.
4. Let's eliminate people since a computer could perform better.

One could say that cases such as US1549, QF32, U232, AA96 or others, where the human contribution is clear, are so scarce that it pays to remove people from the system, because the chances of an accident resulting from human error are higher than the chances of it resulting from an unresponsive, slow or mistaken computer.

Although there are no specific figures to defend the opposite view, many accidents are avoided well before an emergency appears; that is, the potential event is managed before becoming an emergency. A minor and unreported intervention, based even on the suspicion that something is wrong, leads to the situation being solved before it becomes an emergency or an accident.

Major cases like these are scarce, but micro-cases where the process leading to a major event is stopped in an early phase are very frequent. It is not easy, since there are no specific systems for capturing them, but many of them can be captured by asking the right questions of the reporting system databases.

Finding "major events avoided by high situation awareness" – or similar – is difficult because such a category does not exist. Instead, these highly significant cases must be rummaged for among the results of text searches like "automation failure", "malfunctioning sensor", "false warning" and many other options.

Beyond heroic and widely known interventions, removing people from the system would lead to new events that are routinely avoided. The ignorance of the myriad of micro-cases that can be found only through the conscious effort to find them means that any calculation about the benefits of removing people from the system would be based on false inputs.

Exceptional events are not highly informative due to their exceptionality, while common events that can potentially be converted into an emergency or an accident can deliver important information that is ignored.

The human contribution appears only when managing major emergencies. It appears mainly when avoiding them in the early development phases. This is something to keep in mind when a system is designed. Otherwise, the organization will learn the wrong lesson.

Note

1. This question appears in the book *Thirty Seconds to Impact*, as asked by the first officer of BA38 to the captain, Peter Burkill, when he raised the flaps during the final approach.

Bibliography

Abbott, K., Slotte, S. M., & Stimson, D. K. (1996). The interfaces between flightcrews and modern flight deck systems. FAA HF-Team Report.

Air Accidents Investigation Branch (1995). Report 2/95 Report on the incident to Airbus A320-212, G-KMAM London Gatwick Airport on 26 August 1993.

Air Accidents Investigation Branch (2014). Formal Report AAR 1/2010. Report on the accident to Boeing 777-236ER, G-YMMM, at London Heathrow Airport on 17 January 2008.

Australian Transport Safety Bureau (2013). AO-2010-089 Final Investigation In-flight uncontained engine failure Airbus A380-842, VH-OQA overhead Batam Island, Indonesia, 4 November 2010.

Baberg, T. W. (2001). Man-machine-interface in modern transport systems from an aviation safety perspective. *Aerospace Science and Technology*, 5(8), 495–504.

Bainbridge, L. (1983). Ironies of automation. In: *Analysis, Design and Evaluation of Man–Machine Systems 1982* (pp. 129–135). Oxford: Pergamon.

Bennett, K. B., & Flach, J. M. (2011). *Display and Interface Design: Subtle Science, Exact Art*. Boca Raton, FL: CRC Press.

Boeing (2018). Statistical Summary of commercial jet airplane accidents: Worldwide operations: 1959–2017. http://www.boeing.com/news/techissues/pdf/statsum.pdf.

Brooks, R. A. (2002). *Flesh and Machines: How Robots Will Change Us*. New York: Pantheon Books.

Bureau d'Enquêtes et d'Analyses pour la sécurité de l'aviation civile (2010). Report on the accident on 27 November 2008 off the coast of Canet-Plage (66) to the Airbus A320-232 registered D-AXLA operated by XL Airways German.

Bureau d'Enquêtes et d'Analyses pour la sécurité de l'aviation civile (2012). Final Report on the accident on 1st June 2009 to the Airbus A330-203 registered F-GZCP operated by Air France flight AF 447 Rio de Janeiro–Paris.

Burkill, M., & Burkill, P. (2010). *Thirty Seconds to Impact*. Central Milton Keynes: AuthorHouse.

Business Insider (April 2, 2018). If you have a Tesla and use autopilot, please keep your hands on the steering wheel. https://www.businessinsider.com/tesla-autopilot-drivers-keep-hands-on-steering-wheel-2018-4?IR=T.

Degani, A., & Wiener, E. L. (1993). Cockpit checklists: Concepts, design, and use. *Human Factors*, 35(2), 345–359.

Dennett, D. C. (2008). *Kinds of Minds: Toward an Understanding of Consciousness*. New York: Basic Books.

Dismukes, R. K., Berman, B. A., & Loukopoulos, L. (2007). *The Limits of Expertise: Rethinking Pilot Error and the Causes of Airline Accidents*. Aldershot, UK: Ashgate.

Dreyfus, H. L. (1992). *What Computers Still Can't Do: A Critique of Artificial Reason*. Cambridge, MA: MIT Press.

Dreyfus, H., & Dreyfus, S. E. (1986). *Mind Over Machine*. New York: Free Press.

Dutch Safety Board (2010). Number M2009LV0225_01: Crashed during approach, Boeing 737-800, near Amsterdam Schiphol Airport, 25 February 2009.

EASA, CS25 (2018). Certification specifications for large aeroplanes: Amendment 22. https://www.easa.europa.eu/certification-specifications/cs-25-large-aeroplanes.

FAA Federal Aviation Administration (2008). Energy state management aspects of flight deck automation. CAST Commercial Aviation Safety Team, Government Working Group, final report. https://skybrary.aero/bookshelf/books/1581.pdf.

FAA Federal Aviation Administration (2013). Operational use of flight path management systems. Commercial Aviation Safety Team/Flight Deck Automation Working Group. https://www.faa.gov/aircraft/air_cert/design_approvals/human_factors/media/OUFPMS_Report.pdf.

FAA Federal Aviation Administration (2016). Human Factors Design Standard. FAA Human Factors Branch.

Gabinete de Prevençao e Investigaçao de Acidentes com Aeronaves (2001). Accident investigation: final report all engines-out landing due to fuel exhaustion Air Transat Airbus A330-243 marks C-GITS Lajes, Azores, Portugal, 24 August 2001.

Hawkins, J., & Blakeslee, S. (2004). *On Intelligence: How a New Understanding of the Brain Will Lead to the Creation of Truly Intelligent Machines*. New York: Owl Books.

Held, R., & Hein, A. (1963). Movement-produced stimulation in the development of visually guided behavior. *Journal of Comparative and Physiological Psychology*, 56(5), 872.

Kahneman, D., & (2011). *Thinking, Fast and Slow* (Vol. 1). New York: Farrar, Straus and Giroux.

Klein, G. (2013). *Seeing What Others Don't*. New York: Public Affairs.

Komite Nasional Keselamatan Transportasi Republic of Indonesia (2018). Preliminary KNKT.18.10.35.04 aircraft accident investigation report PT. Lion Mentari Airlines Boeing 737-8 (MAX); PK-LQP Tanjung Karawang, West Java Republic of Indonesia, 29 October 2018.

Landman, A., Groen, E. L., Van Paassen, M. M., Bronkhorst, A. W., & Mulder, M. (2017). The influence of surprise on upset recovery performance in airline pilots. *The International Journal of Aerospace Psychology*, 27(1–2), 2–14.

Leveson, N. G. (2004). Role of software in spacecraft accidents. *Journal of Spacecraft and Rockets*, 41(4), 564–575.

Leveson, N. G. (1995). *Safeware: System Safety and Computers* (Vol. 680). Reading, MA: Addison-Wesley.

Leveson, N. (2011). *Engineering a Safer World: Systems Thinking Applied to Safety*. Cambridge, MA: MIT Press.

Maurino, D. E., Reason, J., Johnston, N., & Lee, R. B. (1995). *Beyond Aviation Human Factors: Safety in High Technology Systems*. Aldershot, UK: Aldergate.

Minsky, M. (1988). *Society of Mind*. New York: Simon and Schuster.

Morin, E. (1994). *Introducción al pensamiento complejo*. Barcelona: Gedisa.

National Transportation Safety Board (1973). NTSB/AAR-73/2 American Airlines McDonnell Douglas DC10-10, N103AA near Windsor, Ontario, Canada, June 12, 1972.

National Transportation Safety Board (1973). NTSB-AAR-73-19 Uncontained engine failure, National Airlines, Inc., DC-10-10, N60NA, near Albuquerque, New Mexico, November 3, 1973.

National Transportation Safety Board (1990). NTSB/AAR-90/06 United Airlines Flight 232 McDonnell Douglas DC-I0-10 Sioux Gateway Airport Sioux City, Iowa, July 19, 1989.

National Transportation Safety Board (2010). NTSB/AAR-10/03 Loss of thrust in both engines after encountering a flock of birds and subsequent ditching on the Hudson River US Airways Flight 1549 Airbus A320-214, N106US Weehawken, New Jersey, January 15, 2009.

National Transportation Safety Board (2014). NTSB/AAR-14/01 Descent below visual glidepath and impact with seawall Asiana Airlines Flight 214 Boeing 777-200ER, HL7742 San Francisco, California, July 6, 2013.

Penrose, R. (1990). *The Emperor's New Mind: Concerning Computers, Minds, and the Laws of Physics*. Oxford: Oxford Landmark Science.

Raskin, J. (2000). *The Humane Interface: New Directions for Designing Interactive Systems*. Reading, MA: Addison-Wesley Professional.

Rasmussen, J. (1986). *Information Processing and Human–Machine Interaction: An Approach to Cognitive Engineering, North-Holland Series in System Science and Engineering, 12*. New York: North-Holland.

Reason, J. (2017). *The Human Contribution: Unsafe Acts, Accidents and Heroic Recoveries*. Boca Raton, FL: CRC Press.

Risukhin, V. (2001). *Controlling Pilot Error: Automation*. New York: McGraw-Hill.

SAE (2003). ARP-5364 Human factor considerations in the design of multifunction display systems for civil aircraft. ARP5364.

Sánchez-Alarcos Ballesteros, J. (2007). *Improving Air Safety through Organizational Learning: Consequences of a Technology-Led Model*. Aldershot, UK: Ashgate.

Senate of the United States (2018). Calendar No. 401 115th Congress 2D Session H. R. 4 in the Senate of the United States May 7, 2018. Received; read the first time May 8, 2018. Read the second time and placed on the calendar.

Sowell, T. (1980). *Knowledge and Decisions* (Vol. 10). New York: Basic Books.

Stanton, N. A., Harris, D., Salmon, P. M., Demagalski, J., Marshall, A., Waldmann, T., et al. (2010). Predicting design-induced error in the cockpit. *Journal of Aeronautics, Astronautics and Aviation*, 42(1), 1–10.

Tesla (2018). Full self-driving hardware on all cars. https://www.tesla.com/autopilot.

Transportation Safety Board Canada (1985). Final Report of the Board of Inquiry Investigating the circumstances of an accident involving the Air Canada Boeing 767 aircraft C-GAUN that effected an emergency landing at Gimli, Manitoba on the 23rd day of July, 1983 Commissioner, the Honourable Mr. Justice George H. Lockwood April 1985.

Wall Street Journal (2014). Airbus shifts pilot-training focus to emphasize manual flying. Europe Business, June 19th, 2014.

Walters, J. M., Sumwalt, R. L., & Walters, J. (2000). *Aircraft Accident Analysis: Final Reports*. New York: McGraw-Hill.

Wells, A. T., & Rodrigues, C. C. (2001). *Commercial Aviation Safety* (Vol. 3). New York: McGraw-Hill.

Wheeler, P. H. (2007). Aspects of automation mode confusion. Doctoral dissertation, MIT, Cambridge, MA.

Winograd, T., Flores, F., & Flores, F. F. (1986). *Understanding Computers and Cognition: A New Foundation for Design*. New York: Addison-Wesley.

Wood, S. (2004). Flight crew reliance on automation. CAA Paper, 10.

4

People as Alternative Resources: Feasibility and Requirements

As mentioned in the previous chapter, people are the main alternative resource, but to be useful in that role, behaving adequately, some conditions must be met. These conditions will define a human-suitable environment.

The first question to answer should be about the correct roles for technology and people. To do so, we should assume that the objective is the highest achievable level of functionality, understood as the right mix of efficiency and safety, and since this a high-stakes business, this objective should remain in abnormal or degraded situations.

Furthermore, present automation development guarantees that, as far as everything goes as expected, the operation will be smooth and safe. However, people are asked to give their best in degraded situations; to do so, they will need some resources, not only related to information given when the emergency arises.

Managing adequately the emergency will require some actions to be performed while everything is running normally. One of the best reminders of this principle can be found in the official report of AF-447, which points out – correctly – that the assessment of the situation was difficult for the captain because he was not present when the abnormal events started. By the time he discovered what was happening, it was too late. The commonly used expression "being in the loop" is precisely about that. So, we should examine whether the present practices help to meet that objective and, if not, what is the alternative action.

Common Practice

The most widespread practice, regarding normal operations, has been to bet on technology wherever an improvement in efficiency can be shown. This practice will suggest further issues that will open up some questions related to its adequacy.

Even if the tasks assigned to people are carefully crafted from the ergonomic (strictly physical) point of view, they can lack internal consistency.

Tasks are evaluated only in terms of feasibility – that is, whether the physical features, skills and workload allow the tasks to be performed. In an environment where situation awareness is critical, the information required to keep it is supposed to be only an output of the system.

The possibility of inducing situation awareness by performing specific tasks – other than queries to information systems – is not considered in design. Furthermore, some of the tasks that could guarantee that situation awareness – usually related to manual operation – are not possible in the present environment.

> As an example, manually flying a non-fly-by-wire (FBW) plane implies feeling the weight of the control and the trends of the plane.[1] It can give the pilot indications about trimming, about the center of gravity and about speed. Processing these feelings could lead, for instance, to changing the fuel feeding.
>
> However, the precision required in situations such as flying reduced vertical separation minimum (RVSM) makes manual flight impossible.

This is not the only example. It can be extended to other activities – for instance, take-off calculations. After the Emirates 407 tail strike, Australian authorities included some interesting conclusions in the official report:

> The inadvertent use of erroneous take-off performance data in transport category aircraft had resulted in a significant number of accidents and incidents prior to this accident and, as identified in the ATSB research study, the problem continues to occur. This has been identified by several investigation agencies as a significant safety issue and there have been studies into the factors involved.
>
> The review of the previous research and similar occurrences identified that they were not specific to any particular aircraft type, operator or location, and that the only common factor in all of the events was that the crews attempted to take off with no awareness that the take-off data was not appropriate for the aircraft on that flight.
>
> Although these events differed in how the errors occurred, the outcome and effect on flight safety was similar and as such, this investigation should be viewed in the context of a much larger safety issue.

The "larger safety issue" could be precisely that an automatic calculation, if the value introduced is wrong but within the accepted range, can lead to major consequences. In other words, when an organization or a procedure is driven by efficiency above all, errors in that system will also be efficient. Bigger effects with result from a minor input.

These examples show something that is usually out of the mind of designers: Tasks can be an information source in themselves. However, the most common option is assigning people those tasks that cannot be performed efficiently by technology.

Workflows are defined by procedures, not by internal coherence. Moreover, when someone tries to achieve that internal coherence (e.g. installing elements in a way that allows them to be checked from left to right and from up to down), experience shows that different contingencies often lead to a breakdown of the predesigned order.

Tasks are not assigned to people because they can keep or increase situation awareness, by themselves or together with other tasks or outputs from information systems. They are assigned only because there is no more efficient way available, not because the assignment makes sense.

Since a major part of efficiency is cost driven and depends on technology improvements and economies of scale, the human role is always questioned and under permanent threat from the advancement of technology and the interest in implementing it, sometimes beyond reasonable use.

This practice is far from being perfect but, unfortunately, it has very clear economic sense: If something can be absorbed into technology, its replication is far cheaper than replication in people. So, there is a very strong incentive to evaluate virtually anything as susceptible to being absorbed by technology.

Anyway, it must be admitted that this principle is a kind of generic process, and it has appeared in different ways while sticking to the same principle; that is, dealing with people is a nuisance to be avoided once technology allows it.

Different milestones can be identified in the process, and, perhaps, one of the most relevant nowadays is the prominence of Information Technology and its side-effects. The big jump forward came with FBW and all the related automation. Now, the same principle is working towards the introduction of Artificial Intelligence and Machine Learning, practices that will be analyzed later.

Finally, many technology-prone people may feel the evolution in terms of safety levels could justify this approach, but something is missed that way: The successful advancement of safety levels is not attributable only to technology but to a mixed system, composed of technology and people.

That may sound like trying to attribute merit to people where it rather pertains to technology, but the reality is far from that: Since many human interventions remain invisible – although they are addressed precisely to stop sequences leading to an emergency or a major event – they are not kept in mind at the moment of making decisions about the role of people and technology.

Using people as a kind of "garbage can", where unrelated tasks are stored while there is no cheaper alternative, could mean a heavy toll in the future. Two facts should never be forgotten.

- An eventual disappearance of the human role means losing many apparently trivial actions that are in fact constantly fine-tuning the system and stopping, in early phases, sequences that could lead to major events. So, decreasing the human role still further could lead to a significant increase in accident rates.

  Right now, the rate is so low that even a 900% increase – as happened between 2017 and 2018 – is not worth analyzing since that is statistical noise coming from the extremely low levels.[2] However, that could change.

- Some practices can enjoy a "silent" period after their implementation. However, their effects can appear some years later. The decrease in training practices could be an example of this phenomenon: Amazing accidents – in the sense that there was a huge disproportion between the identified cause and the effect – are relatively new (although some weird cases can be found in the last few decades), and they could be expected to grow.

The hype related to the application of Information Technology – as a representative of a wider process, aimed at decreasing the human role – seems to show an effect, previously mentioned by Kelly:

> It is always tempting, once a miniature system has proved itself useful within a limited range of convenience, to try to extend its range of convenience ... This kind of inflation of miniature systems is not necessarily a bad thing, but it does cause trouble when one fails to recognize that what is reasonably true within a limited range is not necessarily quite so true outside that range.

This easy-to-understand warning applies to the enthusiasm related to some resources, leading people to use them beyond the point where they are effective. The present displacement of people by the increasing use of Information Technology could follow that pattern. Additionally, there is one factor that makes it difficult to recognize it as a problem:

Information Technology's potential, far from slowing down, has been advancing at breakneck pace; hence, many people are convinced it is an unstoppable, unlimited and almighty option.

A sample of this belief can be found in the words of Rodney Brooks, who identified a path: "Very brilliant people have been drawing stripes in the sand, affirming that computers never would trespass these stripes, while computers did it once and again."

It is interesting to note that Brooks does not offer any rationale to support the supposedly unlimited growth of Information Technology or, in his terms, a proof of the inexistence of that stripe, wherever it may be. He simply offers his personal belief about its inexistence. Of course, that belief could be so biased as the "essentialist" position criticized by him (correctly, by the way).

Therefore, Kelly's warning, applied to the evolution of Aviation, should not be disregarded. Removing people from the system or trying to use them as a kind of biological machine could bring about negative consequences due to the loss of important but invisible activities that increase both safety and the operativity of the whole system.

An example in the opposite sense could be shown by the US1549 case: It is one of the few events where nobody denies – especially after running simulator tests – the outstanding performance of pilots. However, few people pay attention to an unknown hero: The plane itself. Once the pilots had managed to keep the systems alive, the Airbus A320 (an FBW plane) was especially suited for that kind of situation. The protections of the system allowed the plane to glide at the optimum rate, without the risk of stalling and making micro-corrections to the actions of the pilots at the flight controls.

So, in the same way that US1549 had an unsung hero, thousands of less dramatic situations happen every day with equally unsung human heroes. Through minor corrections, they ensure that flights finish ordinarily.

Before going further on the present course, more information about these situations is required. Technology will carry on advancing, but not all the potential routes for that advancement are adequate. Hence, establishing some clear criteria to define the correct course is a key requirement.

Alternative Model

An alternative model should start at the opposite end – that is, defining which tasks will require human features and why. Once this is guaranteed, the parts managed by technology should be defined, as well as the kind of control over them, including the requirement for alternative resources. However, defining why a task should be performed by people can be more complex than expected.

It is easy to foresee situations that, due to their complexity, their unexpected nature or being considered an emergency, will require human features to be managed. Probably, there is a general agreement about the convenience of having people to manage these situations.

> An example could be an engine stop during a take-off run. The algorithm to run this situation would be, in theory, very simple: If the engine stops and the speed is below decision speed (V1), then power should be decreased, and brakes should be applied while ensuring that the plane stays on the runway. That would be all.
>
> However, nobody trusts an automatic system enough to allow it to manage an emergency. The system can receive a false input and trigger the sequence at the wrong moment. A similar situation can happen with engine fires, although there are several cases where the pilots have switched off the good engine instead of the burning one.
>
> It must be remarked that, while emergencies are assigned to humans, landings without visibility or maintaining vertical distance in places where this distance is very short are activities assigned to automatic systems.

The major feature of Aviation, never to be ignored, is that any emergency must be solved while flying. Some other fields don't share that feature: Nuclear power plants can afford automatic systems that, once an abnormal parameter has been detected, stop the activity at once. Cars, trains, ships – all of them can stop and try to look for a solution before starting again.

Unfortunately, there is no "pause" button while a plane is flying, and people are called upon in these situations to "fix the pipes without stopping the water flowing". This situation can be found in surgery too, the main difference being in the number of potentially affected people.

This single fact has critical consequences that can be explained in knowledge management terms: Where should the knowledge required to manage this situation be?

That knowledge can be displaced to the top or to technical specialists in some cases, but if the situation must be managed immediately and without external support, that knowledge must be in the place where the event happens.

Some activities can afford a non-answer situation, if the expected outcome is not important. So, it should be a matter of establishing an optimum level for situations managed by the system while establishing other resources – or plainly forgetting those that cannot be managed by the system as designed.

Unfortunately, high-risk activities cannot adopt that practice. That's why the management of emergencies needs humans. This is a point where theere is almost general agreement. However, it requires conditions. The practice of having resources inside a glass box to be broken in case of emergency can work for fire alarms, but not for people.

Therefore, there is still a second class of tasks that should be assigned to humans: those that, despite being easy to automate and, if eliminated, contribute to decreasing the workload but which are situation awareness builders. Defining these tasks and why they should be assigned to people instead of automated is complex and could be a full line of research.

Some examples will be commented on where automated processes obscured hard-to-diagnose problems that, otherwise, could have been easily identified. Despite the need to deepen this subject, some criteria can be already anticipated.

1. Traditionally, the emphasis while working in human factors has been error avoidance. EASA tried to break this trend in terms of human limitations but, at the same time, human capabilities – that is, fields where the error must be avoided but also where something more than error avoidance should be expected.

   This could be explained in different terms: Success and failure are not two opposite poles of a single dimension. Instead, they work in different dimensions. So, the opposite side of error is error avoidance, not necessarily success.

   Working about error avoidance means very often disregarding, when not impeding, the conditions required for success or insight, as Klein called it. He uses an interesting comparison to show the negative effect of emphasizing the error-trapping instead of creating conditions for the insight or success.

A four-character password is less secure than one with eight characters, which is less secure than sixteen characters, but is a one-hundred-character password buying much more than one with sixteen characters? At a certain point, the additional workload swamps the small gain in security.

So, working on error avoidance can mean overloading the system while, at the same time, some key issues remain unattended.

Insight, it appears, can require a deep knowledge of the different actions and, beyond that, a frame of mind composed of flexibility, unobstructed intention to improve knowledge and a constant search for something wrong. That frame of mind would explain why some people can find insight while others with access to the same data do not.

2. Relatedly, the second point should focus on finding the situation awareness builders that can lead to the insight and, hence, successfully solve an unforeseen problem. There are two possible options, and both should be used to define the human role.
   - Helping to set the correct workload level: As a part of common human factors knowledge, the well-known Yerkes–Dodson model shows that performance and arousal reach their optimums at intermediate levels of workload (Figure 4.1).

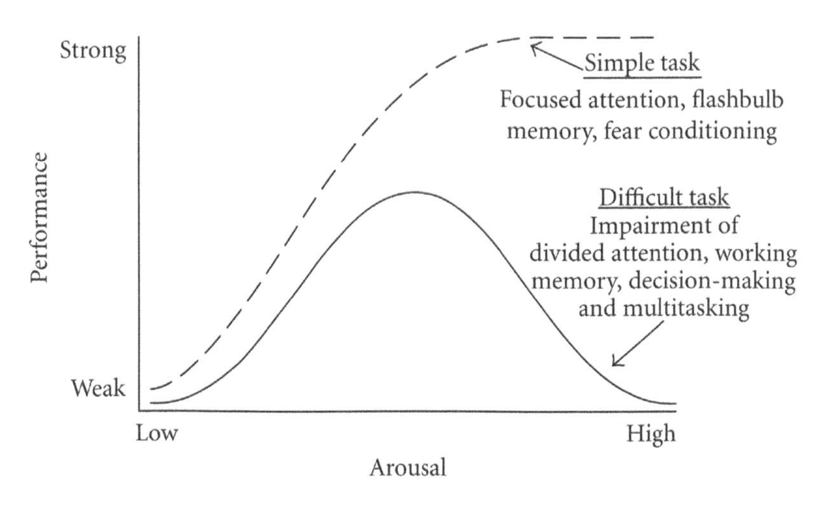

FIGURE 4.1
Yerkes–Dodson model relating performance and arousal. (From https://en.wikipedia.org/wiki/Yerkes–Dodson_law#/media/File:OriginalYerkesDodson.svg [Creative Commons license].)

Unfortunately, some jobs never reach the intermediate level where positive effects happen. They can pass from a high-workload situation, where mistakes happen related to stress, to a low-workload situation, where mistakes happen related to a low arousal level. So, a first clue in adapting designs to human features would not be the "what" but the "when".

If some tasks could be redesigned to be non-intrusive or automated in high-workload phases, it might help. By the same token, before automating a task to be performed in low-workload phases, a second thought would be convenient: Very often, the automated process could steal situational awareness and, hence, remove any chances to manage an unexpected situation.

> Many examples could be offered. For instance, the automation of fuel management can make it harder to know about, in the early phases, any problem related to fuel (Air Transat 236). In the same fashion, automation related to FBW, without feedback in the flight controls (pressure on the hand, normal or abnormal position in the neutral point and keeping both controls in identical position), is clearly detrimental to situational awareness (AF447).

Additionally, many Aviation jobs jump from high to low workloads in standard situations, and abnormal events can be expected to suddenly increase the workload far higher than ordinary values. If "situational awareness stealers" have been avoided, there are many opportunities to properly manage the abnormal event. However, that depends on the nature of the event. So, having the option – not as a default but as an explicit choice – of setting higher automation levels in secondary tasks would be very welcome.

As a rule, and being aware of the exception in the last paragraph, any task that can help to keep the right arousal level could be useful, as far as it is perceived as such – that is, while the operator does not perceive the task as something artificially manipulating the arousal level. For instance, increasing reports or routine checks beyond their functional requirements would not be a valid solution.

So, some useful tasks – even if they could easily be automated – should remain manual to guarantee both arousal and situation awareness. These tasks can be a kind of lifeguard, in the sense that they keep people fit for action and involved.

- The other wide option is related to tasks whose informational contents can let pilots know what is going on. Their value is not related to the arousal level, but to the intrinsic value of the supplied information.

In some cases, the task itself provides valid informative content. In other cases, especially when information systems are involved, the situation is more complex, and this content must be supplied by design.

Perhaps the best attempts in this field have been made by using the ecological interface design (EID) concept, which aims to preserve the chain of meaning through the different levels of the interface. However, despite an excellent theoretical foundation, the practical implementation is not as successful as one might expect.

By trying to mirror the complexity of the "real thing", following the law of requisite variety, designs have been produced that could be good for unknown or unforeseen situations, while common activities could be more complex than in the usual designs. That is, the meaning–efficiency trade is placed at a different point, closer to meaning.

In summary, beyond the interface design, tasks should be designed in a way that the information contained in them and their sequences could add valuable inputs to preserve situation awareness. In some cases, these inputs can come from the intrinsic contents of the task. In some others, they can come from its capacity to keep a balanced workflow.

Very often, designers have a single (wrong) answer for these situation awareness builders: checks. Human Factors knowledge and experience show that humans are better performers than supervisors.

Checks, when imposed, will be performed as part of the discipline of the job but, once perceived as useless or redundant, they could be performed in a reluctant way.

The most typical check is precisely the use of checklists. Degani and Wiener raised a warning related to the redundancy of critical items:

> Although this additional redundancy in the checklist might prevent an item from being missed, overemphasis of many items can degrade the crew's overall checklist performance. Flight crews may degrade the importance of these items if they are being checked several times prior to takeoff.

So, although additional redundancy in checklists, or any kind of check, might prevent an item from being missed, overemphasis degrades the crew's overall performance. Since there is another opportunity to check, this confidence can lead to them missing essential information.

> As an example, in the accident of Spanair JKK5022, which took place in Madrid in August 2008, the pilots missed a critical item on the checklist (flaps setting). Just before the take-off attempt, they voiced the final items, including the intended flaps setting. Obviously, since voicing the item was not done together with a physical check, the item was missed again, leading to a major accident.

We could conclude that redundant checks are not the way to achieve situation awareness, and a better solution, when possible, could come from direct feedback, instead of looking for a specific indicator. Some examples will help to clarify the idea.

- Flight controls do not give feedback through pressure in many modern planes. In the past, a single hand on the controls would be enough to feel tendencies toward the wrong center of gravity, a fuel imbalance or even the proximity to a stall situation.
- Automatic trim used by default disguises that kind of information until an event develops, unless it is specifically checked.
- Feedback from autopilot in flight controls or thrust levers can also let the crew know what is going on, well before a pre-programmed warning level is reached.

An abnormal position in flight controls or thrust levers or any unexpected movement will be instantly detected. Scrolling through the menus to get that same information requires an intentional act. Furthermore, it can be perceived as a useless and bureaucratic task, performed more because of organizational discipline than because of the value of the task.

With the first choice, the pilot will perceive the value of their own contribution, while in the second one, watching will be their main contribution during a major part of the flight – not particularly good for motivation nor arousal. Items can be easily lost when pilots are reduced to watchers.

So, checks should be reduced to a reasonable level rather than increasing them artificially to keep the crew alert or as a side-effect of automation.

Minor tasks, to be performed in low-workload situations but with informative content, would be a better choice. Of course, the philosophy of the manufacturer and the subsequent design is going to define the feasibility of these minor tasks.

> An example related to fuel management: Fuel is critical for a couple of obvious reasons – the first one related to keeping the engines running and the second one because the weight of the fuel must be managed to set the center of gravity in the right place.
>
> A manufacturer could choose a simple system with a small number of tanks and valves to be handled by the pilot, while another could choose a much more complex system with many tanks and valves and whose complexity requires it to be automated.
>
> The second option would probably add some precision to the settings of the center of gravity, but at the same time it would be hard to use manually and, if so, it could be prone to error. So, that design leads directly to automation, with all the linked advantages and inconveniences.
>
> The first option would lose some precision, but the task of manually changing the tanks would provide information, not only about the fuel amounts but about how the plane is flying.

Simple, relevant tasks performed manually give more information than complex tasks that are automated and routinely checked on a specific menu on a screen.

What is better? From the human point of view, the simplified manual tasks are clearly superior to the complex automated way.

Does the difference in efficiency justify complex automation? If so, perhaps a semiautomatic option could be designed to ensure efficiency and, at the same time, avoid losing situation awareness.

Finally, the human role could be filled with minor default tasks without an efficient alternative resource, as far as these minor tasks are compatible with the main ones.

Therefore, as mentioned, the main point here is that task design cannot work with the same rationale as a garbage can, automating everything that can be automated, assigning to the human operator the remaining parts and covering the tasks with warnings, procedures and nice interfaces to avoid human error. For a long time, this has been a common way to deal with human operators.

The design principle of putting human features at the center should be upheld but, to make things more complex, the evolution leads to a permanent task review. Thus, new situations arise where human features were once required.

An example can be found in navigation tasks. Navigating was, in the past, a complex task, far more complex than flying as some specific cases show: So-called Operation Bolero during World War II meant crossing the Atlantic Ocean with the navigational resources of that time and where an expert navigator could perform the task for several planes that were following the one with the navigator.

Nowadays, any plane knows its precise position within a few meters, even if the plane is flying in the middle of an ocean and without an increase in the workload.

Furthermore, there are ever more navigational resources that render obsolete the knowledge that was crucial in the past. So, major technological breakthroughs redefine the human tasks time and again.

Even so, basic principles in the definition of the human role should remain: The human role cannot be defined by default but by clear intentions during the design phase.

Situation awareness has been mentioned in several opportunities as necessary if people are to manage emergencies. The idea of awareness itself is clear enough to suggest what the concept is addressing, but Endsley has developed a full model, including different situation awareness levels (Figure 4.2).

In some ways, the different levels of situation awareness could be related to specific experience: People pass from perceiving scattered elements to putting together a picture of the situation with the most salient elements and, finally, to having a movie where we can see what is coming next.

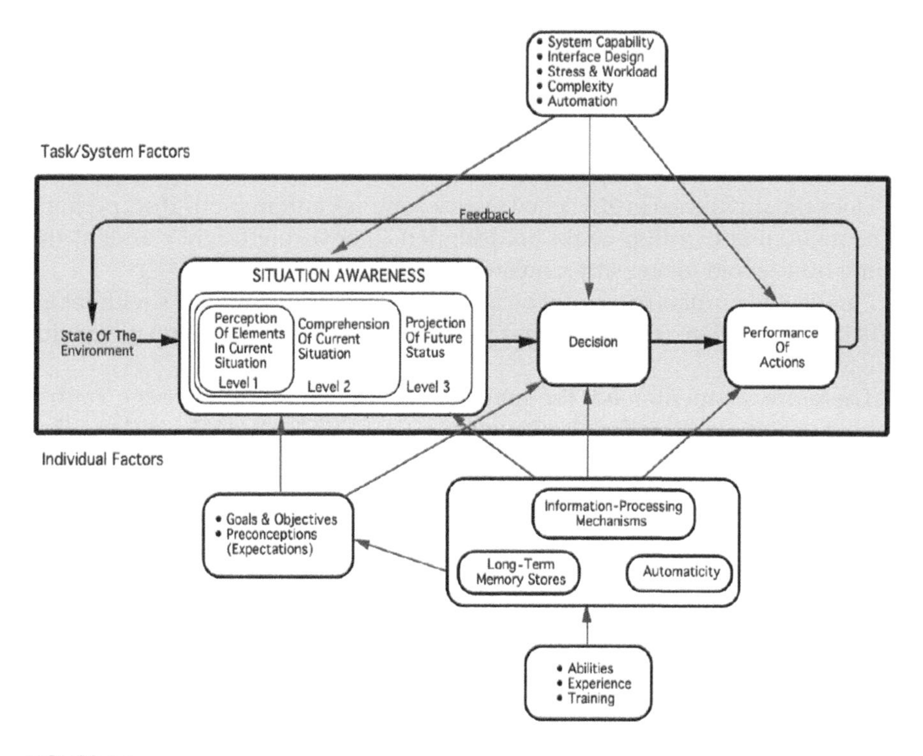

FIGURE 4.2
Situational awareness model.

The quality of this situation awareness would be a basic piece to determine the quality of decisions and, hence, depending on the relevance and risk linked to the task, the level required to perform a specific task can be defined.

This model, as would be expected, has been analyzed and criticized. One of the main criticisms concerns the absence of an explicit reference to meaning. Endsley (2015a) rejected the criticism, pointing out that "even Level 1 data tend to consist of clearly observable, meaningful pieces of information, and higher level comprehension and projection are certainly higher order assessments that are a deep reflection of 'meaning' for that person".

Meaning has a double value, related to motivation and performance:

1. People will be willing to perform a task adequately if there is a clear perception about its usefulness. That perception can come from the task itself – the best option – or from a training process.
2. People will be able to manage unexpected situations if they know the meaning of the performed tasks.

Actually, if people are not given meaning for their actions, they will invent it and, if their guess is wrong, they could deviate from the procedure in unexpected ways.

Many cases can be found where the relevance of a task was ignored by those who should perform it and, therefore, the task was performed in an incomplete or incorrect way:

1. American Airlines 191: The procedure to remove an engine required two separate phases: engine and pylon. Since the process was time-consuming and error-prone due to the double connection–disconnection process, the engineers invented a process where they removed the engine and the pylon in one piece. To do so, they had to put a cart below the engine to hold its weight. If the cart was not in the right place, one element linking the pylon and the wing would suffer since it was not designed to hold, even momentarily, the weight of the engine. The consequence of this misunderstanding was the worst accident in American aviation.
2. Alaska 261: The manufacturer established a checking time for a piece whose access was very difficult. Since this piece was not subject to major pressure, the airline asked for permission to extend the checking times. This permission was warranted and the piece finally failed, leading to a major accident.
3. Aloha 243: Bolts in the upper part of the fuselage had to be checked, but the engineer ignored how critical the task was, even though the problem had been addressed by the manufacturer. Therefore, the maintenance engineer took a sample and the defective parts were missed. The plane would lose a part of the fuselage in midair.

In some other cases, knowing the meaning of the actions is not enough if tasks are performed in an automatic way. A classic model by Jens Rasmussen shows how this source of failures works (Figure 4.3).

The correct behavior can be guaranteed at the knowledge-based level through sensemaking in the task design or linked training. That could be useful for removing, as much as possible, an error source, but many errors will come from automatic actions, performed mechanically, without thinking.

Actually, the most important part of daily behavior is performed at the automatic level, and knowing the relevance of an action does not prevent the mistake. An example will show this.

> A driver that is used to cars with manual gears can have problems driving an automatic car. During the first few minutes of the ride, the driver will probably step on the brake with the left foot while trying to step on an inexistent clutch. The common consequence is violent braking that could lead to an accident.
>
> By the same token, a driver used to automatic gears can drive a manual car for a long time without changing the gear, with potential damage to the engine. The driver might also stop the car simply by braking and, in doing so, will unintendedly stop the engine.

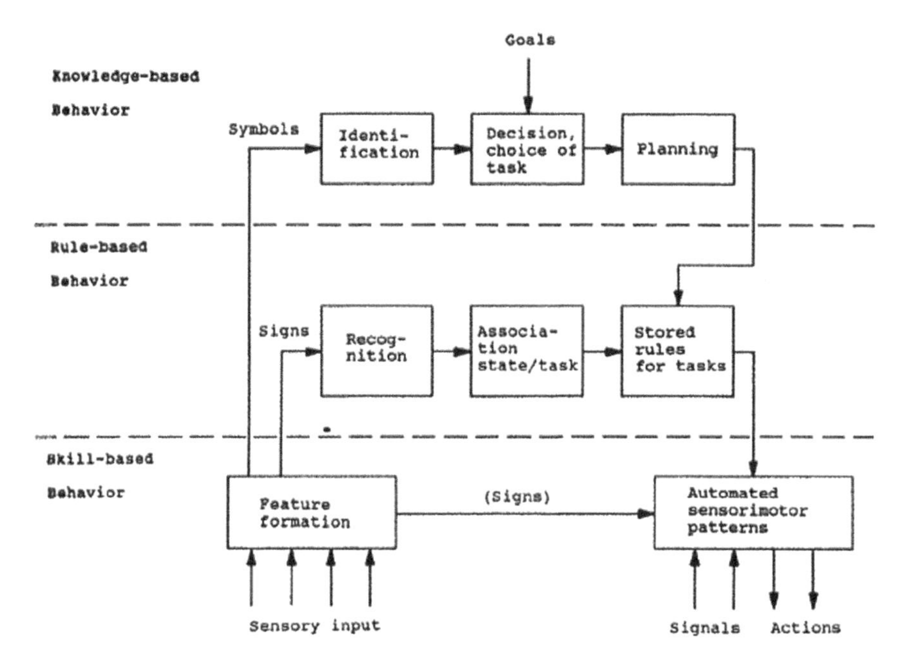

FIGURE 4.3
SRK model. (From Rasmussen, J., *Information Processing and Human-Machine Interaction: An Approach to Cognitive Engineering*, New York: North-Holland, 1986.)

Of course, in both cases the driver knows the procedure, but the automatic action supersedes knowledge and the mistake can appear anyway. As Kahnemann (op. cit.) pointed out in *Thinking, Fast and Slow*, the automatic system cannot be disconnected at will.

Since knowledge is not enough when the action is automatic, three possible solutions can be adopted.

1. Physical design: In the example, the violent braking could be avoided with a narrower brake pedal. Usually, the foot mistakenly hitting the brake does not hit the center of the pedal but the left corner, since the left foot is looking for another inexistent pedal in the left side.
2. Warning: The driver used to automatic cars can be advised of a gear change. This kind of warning is very common even in modern manual cars and it can be useful for both rookies and drivers coming from automatic cars.
3. Procedure: The general procedure is known but, very often, the drivers adopt some "add-ons" to the common procedure. For instance, the driver used to manual cars and conscious of the potential mistake can hide the left foot, bending the leg and placing the foot behind the right leg. In that way, the mechanical action is disturbed and the chances for incorrect action decrease.

The problem would not appear if a common standard was accepted, leading to a single design. The Asiana 214 case has, among its causes, this kind of mistake – that is, two different designs for the same function and a pilot coming from another model and missing an important item. The official report addresses the issue in the following terms:

> He thought the A/T should have automatically advanced the thrust levers upon reaching the MCP-selected airspeed during the accident approach, and he did not understand why that did not occur. Furthermore, he thought the A/T system should have automatically transitioned to TO/GA when the airplane reached minimum airspeed. In that respect, *he believed that the 777 A/T functionality was similar to alpha floor protection on the Airbus A320/321.88.*

It must be noted that performing tasks in an automatic way is not a "lazy" or aimed at avoiding effort. Actually, performance increases where a task becomes automatic, and, in that way, processing resources become free for different and more demanding tasks. Furthermore, focusing consciously on how an already automated activity is done leads to lower performance.

> Some well-known tennis players are said to praise a good hit from the other player, to make that player conscious of it, then less automatic and, then, decreasing the performance of the hit.

So, the solution is not becoming fully aware of automated tasks but managing every one of them with a different set of rules.

The SRK model establishes a hierarchy and its evolution depends on both experience and task design. That will bring specific human contributions and specific mistakes that should be analyzed.

- When we start a new activity, we can expect a high workload related to non-automated tasks – virtually all of them. Mistakes should be expected, but they will come from different origins. Therefore, the solution should be different:
  - Rookie mistakes: Difficulties in separating primary from secondary issues. There is an overflow in the information that cannot be managed, leading to poor decisions.
  - Expert mistakes: The learning of the expert does not start from scratch but from previous knowledge. Some models or assumptions can be applied from former learning but they do not apply to the new environment. That would be the origin of the mistakes committed by the expert.

Therefore, two points should be analyzed: The learning process and the task itself. This will answer a single question:

Which parts of the tasks are expected to become automated and which others should require conscious processing?

Automating a task means liberating processing resources but, at the same time, automating the wrong task could lead to mistakes by applying the wrong automatic process. Every single task becomes automatic through repetition, but in some cases that is an outcome to be avoided because it can lead to errors.

Sometimes, errors are prevented by installing a new device – for example, Emirates installing an acceleration sensor to avoid a potential new Emirates-407 case.

There is an alternative solution: An error-trapping sequence can be designed based on the analysis of the task rather than its repetition, which, as pointed out by Degani and Wiener, develops undue confidence in the second opportunity. Since this same mistake has occurred on different airlines and is related to the same factors, the task could be redesigned to avoid being automated.

So, the first decision related to the task is about which parts should be automated and which others should avoid being automated. Once there, the task could be designed with this goal in mind, using for it different available resources: It can be a different sequence, a specific warning, a device in an uncommon position (like some cars, whose door handle is at the rear, forcing you to look at the rear-view mirror before opening the door) or many other options.

However, as the Australian authorities pointed out in the Emirates-407 report, there is still a major issue. They did not define it, but it's not hard to find it: There us a disproportion between input and output for the person who is performing the task. An apparently trivial task (e.g. inputting a number into a keyboard) can lead to major effects if factors like distraction, fatigue or others converge.

The task, as designed, requires no special skills other than paying attention. Many other tasks will share this feature: not requiring high-level skills or knowledge but only attention, which will fail under fatigue, time constraints or many other environmental situations.

All the tasks, if practiced enough, tend to become automated. If this is acceptable for the specific task, the procedures should be designed to be as intuitive as possible to facilitate the process of becoming automatic.

However, there are tasks where automatic action is not desirable: those where an apparently minor mistake can lead to a major outcome and those that can trigger the wrong automatic actions coming from former experience. If a similar task, performed in the past with a different configuration, can be incorrectly performed because of an automatic and unconscious process, the conflicting parts of the task should be explained in detail.

Of course, that is something to be dealt with during the training process on behalf of the manufacturer, but this is not the only available way: Regulators could enforce standards in the interfaces or the behavior of the systems to avoid these mistakes.

> Some cases have been already mentioned, such as feedback in the flight controls or the existence or not of movement in the thrust levers coming from the autopilot, but perhaps the most notorious case of a different

> standard is of the artificial horizons manufactured in Russia and the confusion among pilots used to a different format.
>
> Some accidents can be traced to a confusion coming from this origin, although major regulators have established clearly how the artificial horizon should be built.
>
> AMC 25.1303(b)(5) 1.3: The display should take the form of an artificial horizon line, which moves relative to a fixed reference aeroplane symbol so as to indicate the position of the true horizon.

It should be noted that this rule was written long ago. However, Russian manufacturers had their own regulations, and, in some places, the mix was hard to avoid, especially in countries flying Russian military aircrafts and with civil pilots trained in this environment.

These differences, then, should not be ignored. The common policy of speaking about one's own system, ignoring the remaining options, can lead to major mistakes. If, as part of a commercial policy, someone wants to avoid mentioning competitors, emphasizing how their own system works in the conflicting parts should be enough.

Ignorance of how a system works, supposing it must be like other known systems, is an important and avoidable source of errors. Cases like the aforementioned G-KMAM or Asiana 214 speak for themselves about risks that should be managed.

On the side of the regulators, the behavior shown at setting standards for artificial horizons is a good example: Standards should be set for other systems beyond generic requirements, even if manufacturers call them by different names while they have similar functions.

In summary, a knowledge-driven task must be analyzed before designing the learning process: The task will evolve, becoming more automated, and this evolution can lead to mistakes. If a negative outcome can be anticipated, the full automation of the task should be avoided by introducing error-trapping elements – that is, different sequences in similar systems.

On the other hand, if automation of the task is desired, procedures should be defined to facilitate it while situation awareness is built.

Finally, issues leading to negative transfer should not be dismissed by manufacturers who are concerned only with their own systems nor by regulators, who should set common interface standards for them.

The next level to analyze should be the *rules*-based level. It is an extremely important level since, very often, it becomes the door to access the other two levels, *knowledge* and *skills*.

Simple procedures become skills, while at a deeper level the learning process should go beyond the know-how and give inputs in the field of know-why – that is, knowledge – so that users are aware of what the system is doing. Otherwise, unexpected events will not have a solution beyond procedure, if it exists. So, procedures should be designed to prevent error, they should be intuitive and they should have actions and sequences that build knowledge.

Finally, the last level to analyze should be the skill-based level. Since this level deals with automated actions, knowledge about the correct action is not enough. Even knowing what action to perform, mistakes can appear. So, the design of skill-based tasks – from the beginning of or after a learning process – should consider former tasks that could interfere through established habits.

If that is not possible, the second option should be the contention – that is, avoiding the consequences of a mistake. Commonality together with cross-crew rating is error inducing, as many incidents have clearly shown. A pilot in a rush can forget which plane he is flying and behave in a way that is only appropriate for a plane 100 tonnes lighter.

So, two different solutions appear for the error-inducing situation if, at the same time, manufacturers and operators want to keep the savings obtained through commonality and cross-crew ratings.

1. Prevention through changes in the procedure: These changes would not be functionally required. Furthermore, minor changes should be performed in the design to prevent the common procedure from working. However, by enforcing a different sequence, the mistake could be instantly detected.
2. Contention through specific warnings: This is the solution applied in the example. An acceleration sensor was installed to warn the pilot that the acceleration values are lower than expected.

In summary, tasks should not be defined by default but by looking for the best situation awareness level and by preventing and containing potential mistakes.

Once this part of the design is prepared, when tasks become informative and help to maintain the correct workload, and when all the automated tasks (or those prone to becoming automated) no longer lead to error, the moment to define which and how other tasks should be automated has arrived.

Notes

1. Boeing kept this feature (kinesthetic feeling) in Boeing FBW planes by keeping hydraulic feedback.
2. The year 2017 was exceptional. Some people called it t "the safest year in Commercial Aviation history", despite some others spoke about luck: However, figures for 2018 were far worse. Actually, both (2017 and 2018) can be read as noise – that is, non-significant movements around a general trend.

Bibliography

Abbott, K., Slotte, S. M., & Stimson, D. K. (1996). The interfaces between flightcrews and modern flight deck systems. FAA HF-Team Report.

Air Accidents Investigation Branch (1995). *Report 2/95: Report on the incident to Airbus A320-212, G-KMAM London Gatwick Airport on 26 August 1993.*

Air Force, U.S.A. (2008). *Air Force Human Systems Integration Handbook.* Air Force 711 Human Performance Wing, Directorate of Human Performance Integration, Human Performance Optimization Division.

Ashby, W. R. (1991). Requisite variety and its implications for the control of complex systems. In: *Facets of Systems Science* (pp. 405–417). Boston, MA: Springer.

Ashby, W. R., & Goldstein, J. (2011). Variety, constraint, and the law of requisite variety. *Emergence, Complexity and Organization,* 13(1/2), 190.

Australian Transport Safety Bureau (2011). Investigation number: AO-2009-012. Tailstrike and runway overrun: Airbus A340-541, A6-ERG, Melbourne Airport, Victoria, 20 March 2009. https://www.atsb.gov.au/media/5773945/ao-2009-012_final-report.pdf.

Aviation Safety Network (2019). Fatal airliner hull-loss accidents. https://aviation-safety.net/statistics/period/stats.php?cat=A1.

Bainbridge, L. (1982). Ironies of automation. In: *Analysis, Design and Evaluation of Man–Machine Systems 1982* (pp. 129–135). Oxford: Pergamon.

Bennett, K. B., & Flach, J. M. (2011). *Display and Interface Design: Subtle Science, Exact Art.* Boca Raton, FL: CRC Press.

Booher, H. R. (1991). *MANPRINT: Implications for product design and manufacture. International Journal of Industrial Ergonomics* 7 (3), 197–206.

Brooks, R. A. (2002). *Flesh and Machines: How Robots Will Change Us.* New York: Pantheon Books.

Buckton, H. (2016). *Friendly Invasion: Memories of Operation Bolero 1942–1945.* Gloucestershire, UK: The History Press.

Comisión de Investigación de Accidentes e Incidentes de Aviación Civil (2011). A-032/2008 Accidente ocurrido a la aeronave McDonnell Douglas DC-9-82 (MD-82), matrícula EC-HFP, operada por la compañía Spanair, en el aeropuerto de Barajas el 20 de agosto de 2008.

Crandall, B., Klein, G., & Hoffman, R. R. (2006). *Working Minds: A Practioner's Guide to Cognitive Task Analysis.* Cambridge, MA: Bradford.

Degani, A., & Wiener, E. L. (1993). Cockpit checklists: Concepts, design, and use. *Human Factors,* 35(2), 345–359.

Dekker, S. (2017). *The Field Guide to Understanding "Human Error".* Boca Raton, FL: CRC Press.

Dennett, D. C. (2008). *Kinds of Minds: Toward an Understanding of Consciousness.* New York: Basic Books.

Diaper, D., & Stanton, N. (Eds.). (2003). *The Handbook of Task Analysis for Human–Computer Interaction.* Boca Raton, FL: CRC Press.

Dismukes, R. K., Berman, B. A., & Loukopoulos, L. (2007). *The Limits of Expertise: Rethinking Pilot Error and the Causes of Airline Accidents.* Aldershot, UK: Ashgate.

DoD. (2012). Department of Defense Design Criteria Standard: Human Engineering (MIL-STD-1472G). DoD, Washington, DC.

Dreyfus, H., Dreyfus, S. E., & Athanasiou, T. (2000). *Mind Over Machine*. Simon and Schuster.

Dreyfus, H. L. (1992). *What Computers Still Can't Do: A Critique of Artificial Reason*. Cambridge, MA: MIT Press.

Dutch Safety Board (2010). Number M2009LV0225_01 Crashed during approach, Boeing 737-800, near Amsterdam Schiphol Airport, 25 February 2009.

EASA (2019). EASA preliminary safety review 2018: Commercial air transport operations. https://www.easa.europa.eu/sites/default/files/dfu/Info%20Graphic%20for%202018%20Preliminary%20Safety%20Review%2031%20Dec.pdf.

Egyptian Ministry of Civil Aviation (2004). Final report of the accident investigation Flash Airlines Flight 604, January 3, 2004, Boeing 737-300 SU-ZCF, Red See off Sharm-El-Sheik, Egypt.

Endsley, M. R. (2015a). Situation awareness misconceptions and misunderstandings. *Journal of Cognitive Engineering and Decision Making*, 9(1), 4–32.

Endsley, M. R. (2015b). Final reflections: Situation awareness models and measures. *Journal of Cognitive Engineering and Decision Making*, 9(1), 101–111.

Federal Aviation Administration (2008). Energy state management aspects of flight deck automation. CAST Commercial Aviation Safety Team, Government Working Group, final report.

Federal Aviation Administration (2013). Final report: Operational use of flight path management systems. PARC/CAST Flight Deck Automation WG.

Federal Aviation Administration (2016). Human Factors design standard. FAA Human Factors Branch.

Fleming, E., & Pritchett, A. (2016). SRK as a framework for the development of training for effective interaction with multi-level automation. *Cognition, Technology and Work*, 18(3), 511–528.

Gabinete de Prevençao e Investigaçao de Acidentes com Aeronaves (2001). Accident investigation final report: All engines-out landing due to fuel exhaustion Air Transat Airbus A330-243 Marks C-GITS Lajes, Azores, Portugal, 24 August 2001.

Gigerenzer, G. (2007). *Gut Feelings: The Intelligence of the Unconscious*. New York: Penguin.

Gigerenzer, G., & Todd, P. M. (1999). *Simple Heuristics that Make Us Smart*. Oxford: Oxford University Press.

Hawkins, J., & Blakeslee, S. (2004). *On Intelligence: How a New Understanding of the Brain Will Lead to the Creation of Truly Intelligent Machines*. New York: Owl Books.

Kahneman, D., & Egan, P. (2011). *Thinking, Fast and Slow* (Vol. 1). New York: Farrar, Straus and Giroux.

Kelly, G. A. (1955). *The Psychology of Personal Constructs, Volume 1: A Theory of Personality*. New York: W.W. Norton.

Klein, G. (2013). *Seeing What Others Don't*. New York: Public Affairs.

Klein, G. A. (1993). *A Recognition-Primed Decision (RPD) Model of Rapid Decision Making* (pp. 138–147). New York: Ablex.

Leveson, N. (2011). *Engineering a Safer World: Systems Thinking Applied to Safety*. Cambridge, MA: MIT Press.

Liskowsky, D. R., & Seitz, W. W. (2010). *Human Integration Design Handbook*. (pp. 657–671). Washington, DC: NASA.

Mansikka, H., Harris, D., & Virtanen, K. (2017). An input–process–output model of pilot core competencies. *Aviation Psychology and Applied Human Factors, 7*(2), 78–85.
National Research Council (1997). *Aviation Safety and Pilot Control: Understanding and Preventing Unfavorable Pilot–Vehicle Interactions.* Washington, DC: National Academies Press.
National Transportation Safety Board (1979). NTSB/AAR-79/17 American Airlines DC10-10, N110AA Chicago, O'Hare International Airport, Chicago-Illinois, May 25, 1979.
National Transportation Safety Board (1989). NTSB/AAR-89/03 Aloha Airlines Flight 243, Boeing 737-200, N73711, near Maui, Hawaii, April 28, 1988.
National Transportation Safety Board (2003). NTSB/AAR-02/01 Loss of control and impact with Pacific Ocean, Alaska Airlines Flight 261, McDonnell Douglas MD-83, N963AS, About 2.7 Miles North of Anacapa Island, California, January 31, 2000.
National Transportation Safety Board (2010). NTSB/AAR-10/03 Loss of thrust in both engines after encountering a flock of birds and subsequent ditching on the Hudson River US Airways Flight 1549 Airbus A320-214, N106US Weehawken, New Jersey, January 15, 2009.
National Transportation Safety Board (2014). NTSB/AAR-14/01 Descent below visual glidepath and impact with seawall Asiana Airlines Flight 214 Boeing 777-200ER, HL7742. San Francisco, California, July 6, 2013.
Netherlands Aviation Safety Board (1992). Aircraft accident report 92-1 1 El Al Flight 1862 Boeing 747-258F 4X-AXG Bijlmermeer, Amsterdam, October 4, 1992.
Parasuraman, R., Sheridan, T. B., & Wickens, C. D. (2008). Situation awareness, mental workload, and trust in automation: Viable, empirically supported cognitive engineering constructs. *Journal of Cognitive Engineering and Decision Making,* 2(2), 140–160.
Penrose, R. (1990). *The Emperor's New Mind: Concerning Computers, Minds, and the Laws of Physics.* Oxford: Oxford Landmark Science.
Raskin, J. (2000). *The Humane Interface: New Directions for Designing Interactive Systems.* Reading, MA: Addison-Wesley Professional.
Rasmussen, J. (1986). *Information Processing and Human–Machine Interaction: An Approach to Cognitive Engineering, North-Holland Series in System Science and Engineering, 12.* New York: North-Holland.
Rasmussen, J., & Vicente, K. J. (1989). Coping with human errors through system design: Implications for ecological interface design. *International Journal of Man-Machine Studies,* 31(5), 517–534.
Reason, J. (1990). *Human Error.* Cambridge: Cambridge University Press.
Reason, J. (2017). *The Human Contribution: Unsafe Acts, Accidents and Heroic Recoveries.* Boca Raton, FL: CRC Press.
SAE (2003). Human Factor considerations in the design of multifunction display systems for civil aircraft. ARP5364.
Sánchez-Alarcos Ballesteros, J. (2007). *Improving Air Safety through Organizational Learning: Consequences of a Technology-Led Model.* Aldershot, UK: Ashgate.
Sowell, T. (1980). *Knowledge and Decisions* (Vol. 10). New York: Basic Books.
Stanton, N. A., Harris, D., Salmon, P. M., Demagalski, J., Marshall, A., Waldmann, T., et al. (2010). Predicting design-induced error in the cockpit. *Journal of Aeronautics, Astronautics and Aviation,* 42(1), 1–10.

Vicente, K. J. (2002). Ecological interface design: Progress and challenges. *Human Factors*, 44(1), 62–78.
Vicente, K. J., & Rasmussen, J. (1992). Ecological interface design: Theoretical foundations. *IEEE Transactions on Systems, Man, and Cybernetics*, 22(4), 589–606.
Walters, J. M., Sumwalt, R. L., & Walters, J. (2000). *Aircraft Accident Analysis: Final Reports*. New York: McGraw-Hill.
Wheeler, P. H. (2007). *Aspects of Automation Mode Confusion*. Doctoral dissertation, MIT, Cambridge, MA.
Winograd, T., Flores, F., & Flores, F. F. (1986). *Understanding Computers and Cognition: A New Foundation for Design*. New York: Addison-Wesley.
Wood, S. (2004). Flight crew reliance on automation. CAA Paper, 10.

5

Organizational Learning in Air Safety: The Role of the Different Stakeholders

Organizational Learning about air safety is not limited to a single organization or a single category since there are many stakeholders: Every state must care for the safety of its citizens. States issue different regulations, by themselves or on behalf of international organizations, while supervising whether or not these regulations are met and how.

Many stakeholders are affected: Airports, Air Navigation Services, security, whatever they are managed by the state or in private hands, manufacturers, operators, individual professionals and, above all, the users.

In some ways, everything is related to every other thing, but this statement does not help to clarify how everything works. So, for the purposes of this book, the analysis has been restricted to the role of manufacturers, operators and regulators and their impact on the learning process. That does not mean disregarding the importance of actors such as airports, ground handling, ATC services or even interest groups such as IATA or IFALPA, but that importance deserves a separate analysis.

At the same time, this analysis partially leaves out a main stakeholder: the user. The different states, through regulatory and supervisory activities, care for the user, and to do so they must be a reliable caretaker. However, users by themselves miss the basic information necessary to maintain an informed behavior. This fact and its eventual effects will be included in the analysis.

Within these limits, two parts will be analyzed: actors and dynamics. The actors – manufacturers, operators and regulators – manage a whole field in a fierce competitive environment, while users seem to be excluded until a major accident occurs and the answer from the system is not satisfactory.

It is surprising to find users excluded from the system, receiving limited and often edulcorated information while being offered the mantra "Aviation is safer than any other transportation system" – asking for blind faith, supported by partial information. Even marketing practices show how this exclusion works:

Low-cost companies have begun to engage in practices that are already present in other operators. They have passed from offering the passengers different perks, making the airline attractive, to introducing artificial nuisances to be avoided by paying extra fees.

Therefore, common passengers are asked to pay more to avoid the long lines at check-in counters or security controls (instead of improving installations and serving them with more people), they are asked to pay more for a change of seat (something that could be done for free in the past) and they are asked to pay to be allowed to keep their hand luggage in the cabin (even if it is within the size and weight limits). These and other current practices show that, unlike many other markets, airlines often don't try to attract passengers – at least, those that fly in economy class, who are by far the major bulk of passengers. Supposedly, their choices are limited to price and the convenience of schedules for their destinations.

Once this is established, operators will fight heavily among them for the best slots and, if they get them, passengers "will come". No more worries about what they want or about keeping them informed.

That's why, contrary to the most obvious choice, passengers are analyzed only in a limited way. With the present management practices, they should be considered a *dichotomic stakeholder* – that is, someone who can flee from a company, from a plane model or from flying but, if they remain, few other options appear for them in the current environment.

Therefore, this chapter will deal with the dynamic of regulations, how they are crafted and the positive and negative impacts that this dynamic produces in the behavior of manufacturers and operators.

A Single Keyword

Usually the risk concept is managed in statistical terms; that is, risk is understood as the product of severity and probability, and the results of this single operation define whether a risk is affordable or not.

The definition should not have anything to criticize, unless someone tries to bend the numbers toward a favorable outcome or there is an honest error; that is, unknown or unforeseen situations show that the estimation was wrong. Even so, something is missing.

Luhmann uses the *disaster threshold* concept to show a relevant parameter that is not included in the statistical analysis of a risk.

> One accepts the results of such a calculation, if at all, only when it does not touch the threshold beyond which a (however unlikely) misfortune would be experienced as a disaster.

Obviously, the potential severity of an air accident can be qualified as a disaster and, as such, a statistics-based risk model is not qualified to managed it. It is suitable for preventing operating risks but inadequate to manage the public effect of assumed and non-communicated risks.

Behind every decision or rule that affects risk, there is a parameter whose nature is not technical nor operational but social or organizational. It's a single word: *acceptability*.

It can be managed in different ways:

- The most obvious way is not accepting risks that common users might consider unacceptable. However, this would be a mobile target and, worse, it can be seriously affected by the influence of media on lay people; that is, some situations can easily escape the control of the main stakeholders.
- A less obvious way is the education of users, showing why some practices are safer than they appear and reaching general acceptance *before* a major event makes the assumed risk visible.
- The last and still less obvious way is not to let certain practices be widely known, even if they are not formally hidden. Very often, this seems to be the chosen way, assuming that the risk is implicit in the disaster threshold concept.

The average passenger might be surprised by the fact that operators and manufacturers compete on price, routes, schedules and comfort but never on safety. Safety is expelled from the competition field with a single rationale: If it (whatever "it" means to an operator, a manufacturer, a pilot, etc.) flies, it is safe. Otherwise, it would be prevented from flying by the regulators.

So, regulators act on behalf of users, but to do so they must keep their credibility intact. Some national authorities do not enjoy that credibility – despite being in the International Civil Aviation Organization (ICAO) – while others have been forced to overhaul their practices after major events.

> Two of these major events affected FAA, a national regulator but also a worldwide reference, and the effectiveness of their surveillance on operators and manufacturers was questioned.
>
> The first one is TK981. Despite it happening in Paris, it made evident a design fault in a plane manufactured in the United States and, hence, under the FAA surveillance. After the accident, it became known that a close call happened 2 years before and, still worse, the defect was known during the design phase but was ignored.
>
> The second one is Alaska-261. An element of the plane that was hard to maintain – due to poor design regarding accessibility – had its checks delayed further and further, with FAA acceptance, until a major accident made it evident that it should have been checked more frequently.

Of course, safety is important to users, and they would not fly, at least not consciously, any unsafe plane or operator, but the lack of information, where regulators ask indirectly for blind faith, drives many of them to make erroneous safety-related decisions.

An average user will decide which operator to choose based on facts such as the age of their planes, comfort, cleanliness or presentation.

It would be unfair to qualify these criteria as irrational, although they are fully unrelated to critical factors such as maintenance quality or crew training. However, if an operator is flying visibly old planes and has a poor image regarding cleanliness or presentation, the user could reasonably expect this operator to also save money in serious and harder-to-see issues.

The other extreme of the same issue is not so clear: New, well-presented and comfortable planes can speak to a healthy flow of money – unless the operator is trying to cover its tracks with these external elements – but that does not inform users about management practices. Cultural elements such as authority gradients can be extremely important for safety, but the user does not know about that.

Recruiting and hiring practices and whether or not operators are driven only by competence are also black holes that can be covered up by a comfortable and well-presented plane. Based on the "superior rationality of the experts", a valid license is supposed to guarantee an acceptable safety level, even though nobody would dare to affirm that the safety level is identical whoever the professional may be.

So, using the pilots as the most obvious example, an operator can choose between "cheap licenses" or "good pilots", but this choice, with obvious consequences in safety level, is unknown to the user.

Despite these and other issues, the safety mantra is repeated time and again to the users: "If it flies, it is safe. Otherwise, it would not fly or would be blacklisted by the main agencies." That limited approach regarding information might be the most common, but even so, it is far from being the only one.

Some local agencies can choose and adapt to international standards or keep their own, at least for local activity, as with China, which certifies passenger transportation planes produced in China to fly only in China.

So, an informed passenger can see that, in some places, a local flight with a local operator is dangerous business. Some people would prefer more reliable operators for local flights in some countries. That is not always possible, not only due to availability but to the general code-sharing practice. This could drive passengers to unintentionally fly with an otherwise undesirable operator.

In summary, people fly because they trust that the system is able to transport them between two points in a safe and fast way. However, people are not given information about how this system works beyond a few edulcorated statements. Users don't have a voice in it, except choosing – when possible – where to use their wallets, and that leads to a peculiar system.

Users trust or don't trust. No grey areas or conditional trust exist since the information to set up the right conditions is missing. So, the system tries to keep users in a bubble of unconditional faith. As such, the bubble could explode, especially in a world of social networks and trending topics, spreading true and false information at light speed.

Keeping acceptability as a major factor is as critical as always, but there are new dangers if, from the system, the main actors insist on preserving unconditional and uninformed faith.

Certainly, an environment where decisions are not externally checked is preferable for the main players but, at the same time, people are nowadays more susceptible than ever to massive information and disinformation, from different channels.

> A good and recent example of this impact can be seen in relation to the different accident ratios in 2017 and 2018. Ratios are so low that a single major accident can have an impact on them, and exactly that happened: The ratio passed from 0.06 fatal accidents in 1 million flights in 2017 to 0.36 fatal accidents in 1 million flights in 2018. The absolute numbers were negligible and, hence, no one could say that there was a real loss in terms of safety level. However, some media reported that the accident rate had grown by 900%, and all the affected parties had to defend themselves from a statement that, despite being ridiculous, had a public effect.
>
> Some other threats have been commented on where an eventual negative headline for Aviation could not be simply disregarded. They could be legitimated precisely by the "inner circle" approach to some critical issues.

Therefore, the comfortable approach coming from poorly informed users could hide a heavy toll to be paid in the short term. That toll would be the loss of acceptability and its consequences.

A few *black swans*, not far in time from each other, could trigger a loss of acceptability, especially if they come from assumed risks. Since, by definition, black swans cannot be foreseen, acceptability should be managed as a relevant parameter on its own, not as a collateral effect of statistics.

Right now, virtually anything can become an unmanageable "trending topic" and, once established, fighting against it can backfire, much like the so-called "Barbra Streisand effect".[1] So, there is an organizational risk that cannot be entrusted to statistics and technical rationality. The organizational risk is not there.

The Manufacturers' World

As a general principle, we could say that Aviation is a global business and manufacturers try to compete worldwide with their products. If someone wants general acceptance, the right way is to be certified by one of the major agencies – that is, EASA for Europe or FAA for the United States. Furthermore, there is a constant effort to make rules coming from both sources compatible when not identical. Hence, getting approval from one of the regulators means you are at least halfway to getting approval from the other.

Many other local authorities consider acceptance by these major agencies as a guarantee, and planes allowed to fly in the United States or Europe are automatically accepted, but the opposite way does not work, unless there are specific agreements with a civil aviation authority. These agreements imply a careful check of the local regulations and supervision procedures and they are hard to get, not only for functional reasons but for non-explicit competitive reasons.

Manufacturers' design and development costs are so high that they always try to become globally certified rather than limiting sales to the country where they produce their planes. The exceptions to this general rule appear in two cases:

1. Light and low-cost planes with low development costs that are intended to be used only in one country. This is common with microlights or other light planes.
2. Big planes to be used in big countries whose air traffic volume could justify a design only for that country. That could be an option in China,[2] India and, on minor scale, Russia. Other big countries that are less populated, such as Australia or Canada, would not be a workable option because of the market size.

Out of these minor options, big plane manufacturers try to reach the global market, and that means a slow and costly certification process aimed at satisfying the requirements of both major regulators.

As well as being slow, the process is also uncertain and, unless the manufacturer has a special relationship with the regulator – as Boeing enjoys with FAA and Airbus with EASA – many subjective items can be read in unfavorable terms for the new applicant.

It could be said that there is an "inner circle", and to get inside a new manufacturer must pay a heavy toll. Once trust is established, the process can move faster and with a clearer outcome.

A major consequence of this process is the slowness in the introduction of new products. Once a plane is certified, manufacturers try to introduce minor improvements to avoid a whole new process. Therefore, technology potential evolves faster than the products coming from that technology. However, the process has some "back doors", allowing the introduction of new solutions:

Thus, engines and avionics are certified as separate pieces of the plane. Hence, once they have the certification, they must be accepted for installation in the certified plane, but the new equipment does not require a full certification process as it is an improvement to an already certified plane.

That's why we can find new avionics and new engines in planes certified in 1968, such as the Boeing 737. At the same time, other changes such as introducing new *fly-by-wire* (FBW) controls would mean a new plane and a full certification process. The different pace of technology and product development leads to weird situations in the market: When the Airbus A320 – designed with FBW – appeared, Boeing claimed that the traditional control was better in Commercial Aviation. However, once Boeing had its own FBW planes, that defense was hard to sustain, and the old model appeared outdated, without any other selling point other than price.

Furthermore, the task of the manufacturer does not finish once the plane is certified to fly. They maintain a close relationship with the operators and the regulators to detect any unexpected problem and how it could evolve during the whole life of the plane.

Manufacturers become the advisors on maintenance practices, malfunctions and other findings. The authorities must also be notified about some of these findings. Once they have analyzed the different issues, manufacturers spread the solution to the different users and, if the situation is serious enough, the regulators can issue airworthiness directives, enforcing changes.

This is a high-level process. However, entering into details, it is possible to find cases where this process has failed.

- One of the most remarkable cases is the aforementioned AA191, where the operators asked for a new maintenance process to detach the engine from the wing. They did not get an answer, and different manufacturers started to work using a more efficient process but, in doing so, they did not realize that a piece had been forced far higher than its design parameters allowed. Finally, the piece broke and a major accident took place.
- The Aloha-243 case, where a section of the upper part of the fuselage was lost in flight, happened due to a technical issue previously identified by the manufacturer. As expected, the manufacturer warned the operators to be sure that this was properly checked, but the warning and its critical nature did not reach the engineer who performed the check.
- A failed take-off warning system (TOWS) was not seen as something deserving an airworthiness directive. Instead, after a major accident (Northwest 255), a recommendation was sent to present users asking for a daily check. Some 21 years later, another accident (Spanair 5022) related to a failed TOWS occurred.
- Perhaps the best-known case of a questionable relationship among manufacturers and regulators was discovered after the TK981 accident, where a design failure – flagged before the plane took off due to a close call – seemed to deserve an airworthiness directive. Instead, it was solved with the previously mentioned "gentlemen agreement".

In some ways, the manufacturers are caught in the middle in a high-stakes business. They must win the approval of the regulators, who happen to be the fiduciaries of the users, and to honor that role they must be very careful in the surveillance of safety.

At the same time, they must be flexible enough to ease both the innovation and profitability of the activity. On the other hand, operators want aircraft to be as cheap as possible in every dimension: price, operation, maintenance and training time and costs.

The manufacturers, in this complex environment, try to play their own ace to attract operators: *commonality*. Training time and costs decrease between similar planes and they need to stock fewer spare parts if they are shared among different planes of the same manufacturer. However, commonality also has a drawback: A major design problem could affect virtually all the planes of a manufacturer. Some hypothetical examples can be offered.

- It is up for discussion whether the kinesthetic information coming from flight controls should be kept or not. Since there is no regulation on this feature, major manufacturers have entirely different practices.
- While Boeing managed to keep that information available, even in FBW planes, Airbus removed it, so the flight stick does not provide kinesthetic information. Cases such as AF447 suggest the convenience of having this information.

  Let's suppose for a moment that a new AF447-like event occurs, and this issue is seen as critical to safety. An airworthiness directive would be issued on it. Airbus, in that hypothetical scenario, would immediately be forced to introduce changes to every single plane manufactured by the company after the A310.

  Boeing kept the feature at the cost of greater mechanical complexity,[3] and a directive such as that would alter the competitive positions, but it is hard to know who would profit from it. What if a "virtual kinesthetic indicator" were to be introduced by Airbus and accepted by the regulators? They could have lower mechanical complexity, and at the same time they could offer something that Boeing offers but at a lower cost.
- Other examples of potential or actual rules and their effects on competition can be found. As already mentioned, ETOPS rules virtually killed four-engine planes since the only advantage they could offer – shorter routes on oceanic flights – disappeared. Four-engine planes are now limited to huge planes – passengers or cargo – that cannot be lifted with only two engines. This situation will remain for as long as the technology does not allow smaller engines to provide enough power to lift large planes and still larger ones in the future. If that happens, we will see planes the size of B747s, A380s or Antonov-225s with only two engines.
- Still more examples can be found: Boeing bet on lithium batteries for its B787, a plane that, as much as possible, changes hydraulic energy into electric energy. Tougher regulations on these batteries could alter its chances in the competition with A350. Beyond that, the B787 could be the origin of a new Boeing family in the same way that the A350 could be the origin of a renewed Airbus family. A major design-related event on either of them could affect many other planes from the same manufacturer.

The *inner circle* model might be preferable for the main actors, as they try to meet the requirements of every member while, at the same time, making the emergence of newcomers more difficult, but the work of this inner circle must be always supported by an acceptability level.

The acceptability level, as already mentioned, is defined by faith rather than transparency: If it flies, it is safe. Otherwise, it would not be allowed to fly.

If an event invited questions of that faith, it would mean a major crisis in the activity. The TK981 case, even at a time where social networks did not exist, was a key reason for the disappearance of McDonnell Douglas, and it seriously damaged FAA's credibility.

Nowadays, it is hard to foresee the consequences of a major event such as a full loss of power in an ETOPS plane for reasons other than those already identified as common, a fire related to lithium batteries after the new regulations or another event like AF447 that could question the whole automation model.

From a strictly technical point of view, a statement as unfortunate as that issued after the fatal Tesla accident should be expected: The chances are so low that no change is required.

Probably, anyone willing to give this or any similar answer would be surprised to find that common people would find it outrageous. Movements that, at the same time, decrease the chances of a negative outcome but increase the severity of a potential one are seen as sensible, if the statistics say so, but the *disaster threshold* coined by Luhmann suggests otherwise.

Perhaps that's why, beyond technical considerations, movements in the competitive markets not only maintain the potential of a major event but increase it through the *tightly coupled organization* effect, as Perrow (2011) called it. So, the organizational risk of statistics-based practices is higher than imagined for the survival of the affected organizations.

Furthermore, some corporate movements can further contribute to the spread of an eventual problem: Until now, two major actors have had the biggest slices of the medium- and long-range planes cake – that is, Airbus and Boeing. Regional aviation and business jets play in a different league, with companies such as Bombardier, Embraer and Dassault and newcomers such as Mitsubishi.

Some movements in the medium range have come from Russia with its Superjet-100 from Ilyushin and China with its COMAC-C919, but the adventure is far from easy: through common training and practices, Boeing and Airbus, especially the latter, have tried to make it as easy as possible to move between different models from the same manufacturer.

Having crews and engineers qualified for different planes through a commonality of systems is a good reason to avoid introducing new planes from other manufacturer.

Hence, buying planes from Ilyushin or COMAC could be a dead-end, only valid for small operators on short or medium routes or, alternatively, for low-cost companies with only one plane type in their whole fleet. Therefore, the opportunities in that market are limited.

Things are different in regional aviation, where things have evolved at a very fast pace, triggered by the U.S. policy on Canadian products.

Bombardier – a Canadian manufacturer – suddenly found there was a barrier to introducing its products to the U.S. market. Since Airbus had factories inside the United States, Bombardier decided to sell its newest model to Airbus, to be manufactured as a U.S. product and, in that way, sidestepping the new barriers coming from U.S. commercial policy.

That meant Airbus entering into a new market, and they even renamed the plane: changing it from Bombardier to Airbus. Boeing saw the danger of this movement and immediately made an agreement with the Brazilian manufacturer Embraer, also entering into the regional market.

So, without intending it, and as a side-effect of the commercial policy of a single country, the United States, both big manufacturers found themselves in the regional market. Therefore, the next steps are easy to foresee: They will try to gain efficiency by spreading commonality to regional planes.

That can decrease the options in this market for any other manufacturer – established or newcomer – and, at the same time, it will increase the potential effects of a design problem.

In a few years, we could see planes from 70 to 700 seats sharing systems. From the point of view of efficiency, it is a perfect solution, but it will increase a danger already present: A major design problem could affect virtually all the planes from one manufacturer.

There would then be a major incentive to blame a clear design flaw on "lack of training" or "human error", as some cases have already shown.

However, by doing so, trust in the "superior rationality of experts" could be seriously affected, especially since it is no longer possible to silence different versions of the same fact.

The Regulators' World

The first question to be answered should be who the regulators are. The best known of them is ICAO, an organization of the United Nations devoted to Aviation.

Since the activities of ICAO affect every single operator and manufacturer, rules from ICAO are basic; that is, they can be met by anyone. So, ICAO recommendations about fatigue, flight time limitations and other things are usually superseded by more advanced regulations in the most developed countries.

In more generic terms, ICAO identifies eight critical safety elements for national regulators:

- Primary aviation legislation
- Specific operating regulations

- State system and functions
- Qualified technical personnel
- Technical guidance, tools and the provision of safety-critical information
- Licensing, certification, authorization and/or approval obligations
- Surveillance obligations
- Resolution of safety issues

Even considering this generic level, supposedly suitable for everyone, recommendations by ICAO are far from being followed in many places, the global implementation of these critical elements are still far from reaching global implementation.

The next level is composed of local regulators who are specialized branches of different governments, but among them there are three major powers: FAA, EASA and the Civil Aviation Administration of China.

The power that these regulators hold beyond their borders comes from a single fact: They behave as the doorkeepers to the biggest markets in the world.

If FAA does not approve a plane or an operator, that plane or that operator won't fly in the United States. The same happens with EASA in Europe and with the Chinese authorities. BCAS India might be in a similar situation to China; that is, it could become a global player, but so far it has been less active than its Chinese counterpart since Indian Civil Aviation manufacturers mainly act as suppliers rather than design their own planes.

FAA and EASA try to harmonize their regulations to the point that they frequently share even the reference number of the rules. SCAA from Russia is not considered part of this powerful Commercial Aviation bloc, despite its size and its developments in Aviation design because of its small market size, especially after the collapse of the Eastern Bloc.

However, its influence is far from trivial: Practices such as using meters instead of feet – globally accepted for measuring flight levels – have led to errors and difficulties in operations. Additionally, Russia, especially during the time of the USSR, was a major designer of military aircraft that have been flown by many pilots all around the world.

Other regulators, despite being very active, are less relevant since they don't have the potential to open major markets. Among them, CAA-UK should be mentioned. Despite being within EASA – at least until so-called Brexit – they publish their own documents, very often more advanced than their EASA counterparts, rather than publishing as UK regulations something previously approved by EASA.

In some ways, EASA must balance the requirements for European countries at different levels of development – as ICAO does for all its members – while CAA-UK does not need to perform that exercise in their internal regulations.

Two other large countries with small populations – a perfect mix to guarantee intense Aviation activity, despite not being a major market – are Canada and Australia.

Both are very active in their publications and Canada can be recognized as a pioneer in fatigue risk management system (FRMS) regulations – in spite of the criticisms that followed after a major fatigue-related incident[4] – while CAA Australia has issued valuable documents about human factors and safety and was one of the first adopters of the Human Factors Analysis and Classification System (HFACS) model.

Many other CAAs adopt a more passive position, adapting and using regulations coming from FAA or EASA. Some of them, as already mentioned, fight for a privileged position, in the sense that their certifications can be automatically recognized by one of the major regulators or both. That is the case, for instance, with JCAB (Japan), whose purpose is to certify planes that would be automatically accepted by FAA.

This is where acceptability comes into play again. In some places the surveillance level is considered acceptable, while in others, even if the same rules apply, the effective surveillance of these rules could be questioned.

So, a foreign operator is expected to meet the safety standards of its country, even if it operates in a different country. By the same token, a local operator in a country whose surveillance habits are not trustable can be rejected where possible.

We saw in the first chapter that Aviation safety has advanced, but this advancement is not the same in every country. Some places still have high accident rates, so the role of regulators is extremely important.

Corruption and lack of resources are a dangerous mix, and very old planes keep flying because there are not enough resources for modern planes, filled with electronics and software. Additionally, these places do not have the resources for a qualified surveillance of the whole system and, when new designs appear, things don't always change for the better.

A modern plane, in theory, should be safer both because it is newer and because it has more advanced systems. However, results can be different from expected if the resources in specific training and maintenance don't reach the required level.

Despite this, even operators with few resources are "invited" to buy new planes by limiting their operations, not only through blacklists but through actions such as the prohibition of flying in *reduced vertical separation minimum* (RVSM)[5] space if they don't have specific technical resources that are usually not present on old planes. That means using lower flight levels and more fuel.

We must remember that regulators behave as experts standing up for passengers. By playing this role, they are putting at stake their own credibility, which could be affected by the outcome of a major event and the subsequent investigation.

Although investigations are assigned to different branches of public administration, not the regulator itself, it is hard to sustain – although every state does it – investigation branches as fully independent entities whose analyses and conclusions are not biased. The interests of the state in the market or some operators – whether they are local manufacturers or flag airlines – can bias some conclusions. Furthermore, with people jumping between different career paths, even personal interests can lead regulators to perform softer-than-required assessments: Nobody wants to make an enemy of a potential future employer.

Different cases could be mentioned where independence appears, at best, dubious. However, experience shows that the main guarantee of uncovering the truth often comes from mutual interests rather than theoretical independence.

Frequently discussed cases such as EgyptAir 990, Los Rodeos, WRZ in Buenos Aires, AF447, TWA800 and others have versions that disagree from the official report. Sometimes, these versions are included in the report as an alternative position while others have become known as an alternative interpretation of the facts. As long as facts are not hidden, some of these different versions, when subject to scrutiny, can be more reliable than a supposedly unique and objective truth from a supposedly independent organization.

The Operators' World

If there is a single business that, by its own nature, is global, that is Aviation. Large countries such as Brazil and Canada have specialized in regional Aviation – something that makes sense since the country of origin can be a good market – but, recently, major manufacturers have taken over, and the Bombardier–Airbus and Embraer–Boeing agreements create a different landscape for the future.

In that future, operators will deal with big manufacturers in regional operations too. That will not be an issue for regional operators who are a part of larger airlines, but it could drive minor players out of business because of their lack of bargaining power, and there are not many manufacturers to choose from.

Whatever the situation regarding bargaining power, operators always have a single requirement: efficiency. However, that single principle does not provide useful guidelines since its definition is different depending on the specific markets that the operator is serving.

Furthermore, some operators are state-owned companies, whose mission is more related to the international promotion of the country than with the airline as a business. Actually, some Middle East companies have been accused of illicit competition because of this fact.

Airlines such as Ryanair, Emirates or Iberia play in different arenas and, perhaps, in different games: While Emirates operate with long stops in Dubai, trying to be an East–West bridge and, at the same time, advocating Dubai as a commercial and touristic destination, Ryanair helps to promote new destinations, often receiving financial incentives to fly to them, while trying to reduce costs or increase profits in every imaginable area. Iberia, instead, has designed a network of South American destinations, trying to become the link between Europe and South America. All of them fly planes, but the sizes, the services, the prices and the destinations are quite different.

At the same time, many airlines are engaged in a game that implies both competition and collaboration through alliances and code-sharing practices. A passenger can buy a ticket with an airline and finish flying with a different airline whose practices and services are very different. Furthermore, which airline performs the flight can depend on the day of the week.

The message that these practices convey to the passenger is very simple: Any airline – at least, those included in the same alliance – is equivalent to any other. Choose your flight, try to get the best price for it and that's all.

Even at the most superficial level, passengers can perceive that this is not the whole truth, since comfort and service can vary among airlines in the same alliance. There are good reasons to suppose that, if there are differences in the most trivial parts, there are probably many more relevant differences in other activities such as recruiting, training or maintenance.

- A few years ago, in an incident nicknamed "nut rage", a high-ranking manager of Korean Airlines became angry with a flight attendant and made the pilot go back to the gate to expel him. The incident had negative consequences for the manager – incidentally, the daughter of the president – but nobody asked about the authority of the pilot to refuse to carry out such a bizarre order and why that authority was not used.
- The same company had a major event in Guam (KAL-801), where one of the findings was related to crew resource management (CRM) and the authority gradient. No one dared to challenge a mistaken captain in the same way that the captain in the nut rage case did not dare to challenge the mistaken manager.
- The Asiana case in San Francisco, among other findings, shows that the pilot that was not comfortable flying an approach manually, but at the same time he was not comfortable saying so.
- The Saudia 163 case is perhaps one of the most shameful accidents in the history of commercial aviation. Among other things, when questioned by a flight attendant about preparing an evacuation, the captain answered, "Sit down."
- On the positive side, the QF32 case is an example of teamwork and deep knowledge, and the same can be said of US 1549.

Many more cases could be mentioned, but these are enough to raise a question: Are all airlines alike, as suggested by these practices, in relation to passengers?

Even among the non-blacklisted, practices differ widely: The boarding process of a low-cost company – and the visible urgency to send the plane off to earn money as fast as possible – is completely different from that of a long-haul carrier; policies about on-board fuel and the freedom of the captain to add extra amounts as a contingency are dramatically different; the pressure related to maintenance issues is also different. All of these are within the regulations.

Experienced and high-quality operators used to have other controls beyond the regulations. For instance, the composition of a team, whether it is a maintenance or operation team, goes beyond the validity of their license and they avoid dangerous mixes, such as putting together an inexperienced captain with an inexperienced first officer.

The emergence of rostering programs makes this informal control more difficult. These programs are designed to manage the complex *sudoku* of flight time limitations, vacations, available aircraft and other factors, and unless some pairings are explicitly forbidden in the system, some undesired pairings could escape this informal control.

Therefore, many differences can be found among operators. Even so, there is an ongoing war among airlines over airport slots because they know that, in a world where users are not well informed about what they are buying, people will choose the most convenient schedule (hence the slots war), the best price – directly or through fidelity programs – and the most comfort.

Safety is out of the competition parameters because the commercial aviation world has been crafted that way. The "If it flies, it is safe" mantra might not be true, but it is successful anyway. Hence, organizational learning and all the associated improvements are addressed to become more and more efficient.

Since safety is out of the competition, that leads to standardization, automation, fewer people and less training.

The "If it flies, it is safe" mantra protects this dynamic, as far as it can be protected – that is, as long as credibility is not damaged by a major event, leading to questions about the role of everyone in the whole system.

Conclusions

For different reasons, there is a close relationship between regulators, manufacturers and operators. As mentioned, some other relevant organizations, such as IATA or IFALPA, were not included in the analysis even though they are stakeholders, albeit on a secondary level, compared with the analyzed organizations.

Not every regulator, manufacturer or operator can be considered part of the inner circle; only the most prominent of them are inside it and some people migrate from time to time among them.

ICAO is a member of the United Nations and, as such, it issues recommendations and rules that must be reflected in the local regulations. That means, usually, the minimum level that must be enforced everywhere.

The major players in the big game are EASA–Airbus, as the regulator and the main manufacturer in Europe, and FAA–Boeing, as the regulator and the main manufacturer in the United States. Although EASA and FAA are local agencies and, hence, they have locally limited authority, they are also the doorkeepers to the biggest Aviation markets and, as such, they could be considered global regulators.

Everyone who wants to access to these markets must go through these agencies. The emergence of large and densely populated countries such as China and India as major new markets could have an impact on future global dynamics, but at the moment there are only minor movements, such as the China-only certification from the Chinese authorities of planes such as COMAC ARJ21.

The situation could change if CAAs from other countries are willing to accept the Chinese standard; hence, they would escape from the two most powerful players right now. If so, the overall balance could change.

The Russian Federal Agency for Air Transport is a special case. Russian civil aircraft still have a strong presence in Eastern European countries, and military aircraft, subject to different regulations, are present in many more countries. Could the Russian agency become a global power, competing with EASA and FAA? Right now, the answer should be negative, but Russian aircraft have something that could drive many countries to accept Russian regulations as their own.

EASA and FAA are agencies ruling for the most developed countries in Aviation. Other countries that stay level with Europe and the United States in Aviation development (Japan, Korea, Singapore, Australia, Canada, etc.) accept the main part of these regulations and, in that way, they avoid the need for huge local regulators.

However, the Aviation world is not limited to rich countries, and safety statistics show a distressing tendency.

Safety ratios have been improving, making us forget the apocalyptic prophecy from the White House Commission on Aviation Safety and Security:

> Given the international nature of aviation, cutting the accident rate is an imperative not just for the United States, but for all countries involved in aviation. Accident rates in some areas of the world exceed those in the U.S. by a factor of ten or more. Boeing projects that unless the global accident rate is reduced, by the year 2015, an airliner will crash somewhere in the world almost weekly.

We have seen that safety has improved, but not everywhere. Actually, as already shown, in some countries, especially in Africa, the trend is worsening, and we should ask about the role of technology and regulations in that effect.

Modern planes are very complex, not only in the material used to build them – about 50% of fiber in the fuselage of the most modern models – but also in their systems, filled with electronics and software.

Many operators in poor countries find that they are unable to maintain, or even know how to use, these planes and, hence, they stay attached to old planes, which are easier to maintain with fewer resources but are, above all else, aging.

Russian engineers, with some modern exceptions trying to compete with Western aviation, such as Superjet, are recognized for their strong construction and wide use of mechanical technology, which they have developed to the highest possible level. For instance, before having access to the Russian fighter MIG-29, many Western specialists were convinced that its performance, competing with the American F16 in maneuverability, could only be justified by FBW controls. Afterward, they would find that the plane had mechanical controls.

Many planes coming from this "Russian school" might not pass the EASA/FAA tests or could be limited to non-commercial operations, but they could still win some markets where airworthiness with low resources is a must or where there are exceptional conditions.

Of course, operators have a clear role since, to fly, they need aircraft and rules allowing them to run a profitable business. Otherwise, Aviation simply would not exist.

In some cases, operators have a bigger role than expected. For instance, the Boeing 747 was born because an airline – Pan Am – asked Boeing for a bigger plane than anything previously known, and Airbus continued to sell its A380 based on a single client, Emirates, whose impressive A380 fleet allows this airline to impose on the manufacturer specific features for brand-new planes.

In some ways, if safety agencies are representative of the passengers from the safety point of view, airlines could be seen too as representative of the passengers from the financial point of view: If they can acquire planes that are cheaper to buy, cheaper to operate and cheaper to maintain, they will decrease their costs and, hence, they will be able to offer lower prices.

Therefore, operators press manufacturers for the most efficient aircraft in every dimension – price, operation and maintenance – while safety agencies must know about the practices of both manufacturers and operators in design, operation and maintenance.

The system is learning about safety through major and even minor events, if the analysis reveals the potential for bigger accidents. However, efficiency objectives can bias these internal analyses. The different players are very careful to avoid entering into a war over safety levels. Users are fed the "If it flies, it is safe" idea because, otherwise, safety agencies would not allow planes to fly.

It has worked for decades, but something is lost beyond the technical side and the wide use of statistics:

The capacity of any uninformed "influencer" to spread information worldwide almost instantly is no longer limited to big media. Anything can become a "trending topic", and information reservoirs are not safe in the world of WikiLeaks.

> TK981 revealed the "gentlemen's agreement" between FAA and McDonnell Douglas. In the same case, the Applegate report about the potential the design flaw in the DC-10 revealed that the manufacturer knew about the problem.
>
> However, at the time, a major accident and a full investigation were required to reveal these facts.
>
> Years later, the disasters of *Challenger and Columbia* also revealed organizational flaws, where managers decided to look elsewhere instead of confronting the safety problems. Again, two disasters and investigations were required.

Right now, any dubious agreement or any improper behavior can appear without a major accident or an official investigation. It can affect the credibility of any safety agency and, by extension, the whole activity. Again, the B737MAX issue is an excellent example.

That's the organizational danger within the paternalistic attitude, where passengers are dealt with as children, easy to please with a handful of sweets. This attitude can be observed even in details as minor as telling adults to adjust their oxygen mask before helping children or other people, but not explaining why such "selfish" behavior is required.[6] Similar situations can be found where passengers are not informed why raising the blinds during takeoff and landing is required, why lights are dimmed, why seat backs should be in the vertical position or tray tables should be folded, why passengers are advised to keep their belt fastened, and so on. All of these minor issues where passengers are expected to follow the procedure, without giving them an idea about why these are sensible things to do, show something: The possibility of an accident must be out of the mind of the passenger. Hence, it's not surprising that common passengers trivialize the possibility, and if, finally, an unexpected event occurs, the surprise could bring equally unexpected reactions.

Well-known accidents such as the midair explosions of the Comet were not avoidable with the knowledge available at the time. It could happen again, and the manufacturer was actually considered not guilty in those cases. Real problems with credibility will appear if a foreseeable event occurs where the information was already available to the "inner circle" and the assumed risks were not explained.

Some cases where planes have crashed because of the confusion of the pilots were attributed to "lack of training", despite this hardly being a valid explanation: Lack of training is a tautology since it means that someone did – or didn't do – something without knowing the effects.

Why were the effects unknown? Were they knowable? If so, why was a person without the right training in the wrong place? "Lack of training" always bring us to poor design, unknowable issues, or organizational issues. However, it is a useful wild card to assign the guilt to the weakest link in the chain rather than addressing other major and uncomfortable issues.

In the same way, with the current amount of air traffic, we might expect a twin plane to fall out of the sky due to a full loss of power during an Oceanic flight. If that happens, will someone say, while hundreds of corpses rest on tables or at the bottom of the ocean, that statistics establish the chances as being below one in a billion,[7] that everything is fine and the show must go on?

Keeping information in the hands of experts is the comfortable choice as long as we are sure that this information remains in the expected place. However, in the present world, nobody can be sure that reserved items won't become widely available, which could have uncontrollable effects. That could change, and the loss of comfort of technical experts should be assumed and anticipated.

The alternative is accepting an organizational risk below the *disaster threshold*, which is nowadays ignored by the main players. While safety people play with statistics, common public don't. The rational model of every group is very different and that should not be disregarded.

Some players, such as Southwest Airlines, have shown wise behavior during a crisis, which, unfortunately, is not the rule but the exception. The company received a big fine from FAA due to their maintenance practices. Instead of hiding themselves, they published their own view about the issue. Furthermore, they published the testimony of the CEO, Herb Kelleher, in the trial.

Something similar could be said of Boeing, in the JAL123 case, where a Boeing 747 crashed after a faulty maintenance process performed by Boeing. During the research, they made mathematical models and cooperated while they knew in advance what they were looking for.

Once the evidence was found it was not hidden. The difference in the behavior and outcomes of the JAL123 and Turkish Airlines 981 cases seems evident. While McDonnell Douglas, by hiding information and making agreements with the regulator, had dug its own grave, Boeing kept thriving in the market. Even though the JAL123 accident remains the biggest accident of all time in Commercial Aviation, affecting only one aircraft.

In summary, technical and statistical risk is managed. Although there are some flaws, it could be added that risk is, in general terms, properly managed. Social and organizational risk are not, since many players are not even conscious of its existence or of the changes in the environment, especially regarding the feasibility of controlling the flow of information.

Notes

1. The so-called "Barbra Streisand effect" occurred when she decided to sue someone that had (inadvertently) published a photograph of her house without her permission. Unfortunately, that action brought undue attention to the picture, which would have otherwise gone unnoticed.
2. The Chinese manufacturer COMAC's first plane was certified to fly only in China. This project allowed them to gain experience for a second and more ambitious project: a plane to be sold worldwide.
3. FBW controls do not link flight controls with the controlled surfaces. Everything passes through a computer. Hence, getting feedback from the control means designing linked flight controls/flight surfaces exactly with that objective, and that is what Boeing did.
4. National Transportation Safety Board. Taxiway overflight, Air Canada Flight 759, Airbus A320-211, C-FKCK, San Francisco, California July 7, 2017 NTSB/AIR-18/01 (https://ntsb.gov/news/events/Documents/DCA17IA148-Abstract.pdf).
5. Under this rule, the vertical separation required for planes flying between 29,000 and 41,000 ft is reduced to 1,000 ft. That increases the number of aircraft that can use that airspace, but at the same time, higher precision in altimetry and autopilot features are required.
6. At cruise level, once the masks are released, passengers have around 15 seconds of consciousness if the mask is not used. Using these few seconds, for instance, to help a child before putting on their own mask means risking a loss of consciousness since the child cannot help in kind. This justifies the encouragement to be "selfish".
7. It could be said that something whose chances are less than one in a billion simply will not happen. When, contrary to that simple principle, it happens, it usually shows that the probability analysis had been extremely optimistic (e.g. the United 232 case).

Bibliography

Abbott, K., Slotte, S. M., & Stimson, D. K. (1996). The interfaces between flightcrews and modern flight deck systems. FAA HF Team report.

Australian Transport Safety Bureau (2013). AO-2010-089 Final investigation: In-flight uncontained engine failure Airbus A380-842, VH-OQA overhead Batam Island, Indonesia, 4 November 2010.

Aviation Safety Network (2007). Turkish Airlines Flight 981. Accident report, accessed November 23, 2018.

Bainbridge, L. (1983). Ironies of automation. In: *Analysis, Design and Evaluation of Man–Machine Systems 1982* (pp. 129–135).Oxford: Pergamon.

Beck, U. (1986). *Risk Society*. Sage.

Bureau d'Enquêtes et d'Analyses pour la sécurité de l'aviation civile (2012). Final report on the accident on 1st June 2009 to the Airbus A330-203 registered F-GZCP operated by Air France flight AF 447 Rio de Janeiro–Paris.

Business Insider (April 2, 2018). If you have a Tesla and use autopilot, please keep your hands on the steering wheel. https://www.businessinsider.com/tesla-autopilot-drivers-keep-hands-on-steering-wheel-2018-4?IR=T.

Comisión de Investigación de Accidentes e Incidentes de Aviación Civil (1977). A-102/1977 y A-103/1977. Accidente ocurrido el 27 de marzo de 1977 a las aeronaves Boeing 747, matrícula PH–BUF de K. L. M. y aeronave Boeing 747, matrícula N736PA de Panam en el Aeropuerto de los Rodeos, Tenerife (Islas Canarias).

Comisión de Investigación de Accidentes e Incidentes de Aviación Civil (2011). A-032/2008. Accidente ocurrido a la aeronave McDonnell Douglas DC-9-82 (MD-82), matrícula EC-HFP, operada por la compañía Spanair, en el aeropuerto de Barajas el 20 de agosto de 2008.

Dekker, S. (2017). *The Field Guide to Understanding Human Error.* Boca Raton, FL: CRC Press.

DoD (2012). Department of Defense design criteria standard: Human engineering (MIL-STD-1472G). Washington, DC: Department of Defense.

Dismukes, R. K., Berman, B. A., & Loukopoulos, L. (2007). *The Limits of Expertise: Rethinking Pilot Error and the Causes of Airline Accidents.* Aldershot, UK: Ashgate.

EASA, CS25 (2018). Certification specifications for large aeroplanes, amendment 22. https://www.easa.europa.eu/certification-specifications/cs-25-large-aeroplanes.

Endsley, M. R. (2015). Final reflections: Situation awareness models and measures. *Journal of Cognitive Engineering and Decision Making*, 9(1), 101–111.

FAA Federal Aviation Administration (1987). Aircraft accident investigation report: Japan Air Lines Co. Ltd. Boeing 747 SR-100, JA8119 Gunma prefecture, Japan August 12, 1985. Translation from "Aircraft accident investigation report on Japan Air Lines JA8119, Boeing 747 SR-100". https://lessonslearned.faa.gov/Japan123/JAL123_Acc_Report.pdf.

FAA (2016). Human Factors design standard. Washington, DC: FAA Human Factors Branch.

Fielder, J. & Birsch, D. (1992). *The DC-10 Case: A Study in Applied Ethics, Technology, and Society.* New York: State University of New York Press.

Fischhoff, B., Lichtenstein, S., Slovic, P., Derby, S. L., & Keeney, R. (1983). *Acceptable Risk.* Cambridge: Cambridge University Press.

Fleming, E., & Pritchett, A. (2016). SRK as a framework for the development of training for effective interaction with multi-level automation. *Cognition, Technology and Work*, 18(3), 511–528.

Hale, A. R., Wilpert, B., & Freitag, M. (Eds.) (1997). *After the Event: From Accident to Organisational Learning.* New York: Elsevier.

Harris, D. (2011). Rule fragmentation in the airworthiness regulations: A human factors perspective. *Aviation Psychology and Applied Human Factors*, 1(2), 75–86.

Hubbard, D. W. (2009). *The Failure of Risk Management: Why It's Broken and How to Fix It.* John Wiley.

International Civil Aviation Organization (2016). Safety report: Universal safety oversight audit programme: Continuous monitoring approach results 1 January 2013 to 31 December 2015. https://www.icao.int/safety/CMAForum/Documents/USOAP_REPORT_2013-2016.pdf.

Johnson, W. G. (1973). *Management Oversight and Risk Tree-MORT (No. DOE/ID/01375-T1; SAN-821-2).* Scoville, ID: Aerojet Nuclear.

Junta de Investigación de Accidentes de Aviación Civil (1999). Id. Prevac 19990831 Aeroparque Jorge Newbery (SABE) Buenos Aires, Boeing 737-204C LV-WRZ Vuelo MJ3142.

Leveson, N. (2011). *Engineering a Safer World: Systems Thinking Applied to Safety.* Cambridge, MA: MIT Press.
Liskowsky, D. R., & Seitz, W. W. (2010). *Human Integration Design Handbook* (pp. 657–671). Washington, DC: NASA.
Luhmann, N. (1993). *Risk: A Sociological Theory.* New York: A. de Gruyter.
Ministry of Transport and Civil Aviation (1955). Report of the Court of Inquiry into the accidents to Comet G-ALYP on 10th January 1954 and Comet G-ALYY on 8th April 1954.
National Transportation Safety Board (1979). NTSB/AAR-79/17 American Airlines DC10-10, N110AA Chicago, O'Hare International Airport, Chicago-Illinois, May 25, 1979.
National Transportation Safety Board (1988). NTSB/AAR-88/05 Northwest Airlines, Inc. McDonnell Douglas DC-9-82, N312RC Detroit Metropolitan Wayne County Airport, Romulus, Michigan, August 16, 1987.
National Transportation Safety Board (1989). NTSB/AAR-89/03 Aloha Airlines Flight 243 Boeing 737-200 N73711, near Maui, Hawaii, April 28, 1988.
National Transportation Safety Board (2000a). NTSB/AAR-00/03 In-flight breakup over the Atlantic Ocean Trans World Airlines Flight 800 Boeing 747-131, N93119 near East Moriches, New York, July 17, 1996.
National Transportation Safety Board (2000b). NTSB/AAR-00/01. Controlled flight into terrain Korean Air Flight 801 Boeing 747-300, HL7468 Nimitz Hill, Guam, August 6, 1997.
National Transportation Safety Board (2002). NTSB/AAB-02/01 Aircraft accident brief EgyptAir Flight 990 Boeing 767-366ER, SU-GAP 60 miles south of Nantucket, Massachusetts, October 31, 1999.
National Transportation Safety Board (2003). NTSB/AAR-02/01 Loss of control and impact with Pacific Ocean Alaska Airlines Flight 261 McDonnell Douglas MD-83, N963AS about 2.7 miles north of Anacapa Island, California, January 31, 2000.
National Transportation Safety Board (2014). NTSB/AAR-14/01 Descent below visual glidepath and impact with seawall Asiana Airlines Flight 214 Boeing 777-200ER, HL7742 San Francisco, California, July 6, 2013.
Perrow, C. (1972). *Complex Organizations: A Critical Essay.* New York: McGraw-Hill.
Perrow, C. (2011). *Normal Accidents: Living with High Risk Technologies-Updated Edition.* Princeton, NJ: Princeton University Press.
Presidency of Civil Aviation Jeddah Saudi Arabia (1980). Aircraft accident report Saudi Arabian Airlines Lockheed L-1011, HZ-AHK Riyadh, Saudi Arabia, August 19th, 1980 taken from http://leonardo-in-flight.nl/PDF/Saudia%20163.pdf.
Raskin, J. (2000). *The Humane Interface: New Directions for Designing Interactive Systems.* Reading, MA: Addison-Wesley Professional.
Rasmussen, J. (1986). *Information Processing and Human–Machine Interaction: An Approach to Cognitive Engineering,* North-Holland Series in System Science and Engineering, *12.* New York: North-Holland.
Reason, J. (1990). *Human Error.* Cambridge: Cambridge University Press.
Reason, J. (1997). *Managing the Risks of Organizational Accidents.* Aldershot, UK: Ashgate.
Sánchez-Alarcos Ballesteros, J. (2007). *Improving Air Safety through Organizational Learning: Consequences of a Technology-Led Model.* Aldershot, UK: Ashgate.

Stanton, N. A., Harris, D., Salmon, P. M., Demagalski, J., Marshall, A., Waldmann, T., et al. (2010). Predicting design-induced error in the cockpit. *Journal of Aeronautics, Astronautics and Aviation*, 42(1), 1–10.

Tesla (2018). Full self-driving hardware on all cars. https://www.tesla.com/autopilot.

Wheeler, P. H. (2007). Aspects of automation mode confusion. Doctoral dissertation, Cambridge, MA: MIT.

White House Commission on Aviation Safety and Security (1997). Final report to President Clinton, February 12, 1997. https://fas.org/irp/threat/212fin~1.html.

Wiegmann, D. A., & Shappell, S. A. (2003). *A Human Error Approach to Aviation Accident Analysis: The Human Factors Analysis and Classification System*. Surrey, UK: Ashgate.

Winograd, T., Flores, F., & Flores, F. F. (1986). *Understanding Computers and Cognition: A New Foundation for Design*. New York: Addison-Wesley.

6

The Engine for Organizational Learning: Where It Is and Where It Should Be

Introduction

Figures for accident rates are good. Hence, the performance of Aviation in safety terms qualifies it as a successful model: The accident rate is very low, especially compared with Aviation a few decades ago. So, many people involved in the improvement process should feel satisfied.

Furthermore, apocalyptic predictions such as having a major accident weekly by 2015, as stated by White House Commission on Aviation Safety and Security, chaired by Al Gore in 1998, did not become a reality despite the increase in air traffic. Hence, should we be worried about air safety improvement and its organizational channels? Shouldn't a "business as usual" approach be enough?

The first answer should be that there is no "business as usual" approach. The learning process has been changing over time, and the present stage has some threats that should be eliminated. The good part is that they can be eliminated without major changes.

Short History of Aviation Learning: From the Dawn of Aviation to the Present

Things have changed since the pioneers of Aviation first appeared. Although there are some people that could be called "pre-pioneers" from before the end of the 19th century – for example, Leonardo da Vinci (15th century), the Spaniard Diego Marín (18th century) or the first French ballooners (18th century) – the most practical starting point comes with the attempted controlled flights with heavier-than-air machines.

As with any other field, a trial-and-error model led the first attempts. However, the error was not always crystal clear. A workable solution might be found, but its success could lead to a change where the solution did not work anymore. In some ways, the evolution was showing that success renders obsolete those factors that made it possible.

One of the most significant cases of a fatal success is the German pilot Otto Lilienthal, who died a few years before the flight of the Wright brothers. Lilienthal worked with gliders, flying downhill using his own weight as a control device. Despite big changes in the design of the machines, the technique is not outdated, since that is basically what modern hang gliders still do.

Lilienthal developed remarkable skill at flying in that fashion, and it encouraged him to build bigger and faster gliders. One of them was too heavy to be controlled by his body weight and he lost control and crashed. Both fact and outcome would appear as expected today, but in Lilienthal's time it was far from evident.

So, a good control design had been developed, but this design had some hard-to-see limits to growth: The kind of control used was not valid for aircrafts beyond a certain weight. At the same time, bigger aircrafts should be stronger to avoid breakage and, hence, heavier. So, a major change was required if people wanted bigger and faster planes.

However, controlling an aircraft with body weight is merely an application of a wider concept: controlling an aircraft by displacing its center of gravity. The technique remains in some microlight planes, which are designed as hang gliders whose weight is far heavier than the pilot; the center of gravity in bigger planes is managed by fuel displacement or selective fuel feeding. This practice has became a common in-flight activity, whatever the size of the plane.

The pioneers kept working, but Aviation was considered a dangerous business. Airframes that were less resistant to damage or less controllable than pretended, together with unreliable and heavy engines, represented a bad mix.

Accidents were common, and it was expected for pilots to not only be skilled at their business but be able to overcome almost any contingency in flight, solving mechanical issues, showing a clinical eye for dangerous weather and, of course, navigating without getting lost.

A new and dangerous field requiring so many advanced skills would require very special people, with a deep knowledge of different disciplines and, at the same time, with psychological and physical features that could support the required skills. So, the availability of these people would be a real bottleneck for the development of Aviation.

World War I became the environment in which Aviation would show itself to be something more than an attractive activity for people with an adventurous spirit. The military value of Aviation boosted the technical development of the activity and, once the war was over, the era of the air mail, aerobatic shows, air races and oceanic flights would begin.

Before World War II, new developments based on rigid balloons would take Aviation to a different level. The Hindenburg airship anticipated modern luxury cruise ships, but the disastrous crash of 1937, caused by filling it with highly flammable hydrogen, would mark the end of an era.

Approximately at the same time (1930s), the U.S. Navy would use the first aircraft carriers, but rather than huge vessels they were airships on which light planes "landed" by making the stall coincident with the position of a hook on the lower part of the carrier (Figure 6.1).

The picture should give you an idea about the skill required to perform the maneuver, without stalling but slow enough to avoid breaking the hook or the plane, not to say the full absence of vertigo while getting in or out of the plane.

World War II, beyond technical developments, would bring a new element: If people involved in Aviation needed to be as exceptionally skilled as these pioneers were, it would mean be a serious shortage of people.

Before beginning bombing campaigns in Europe, so-called Operation Bolero implied crossing the Atlantic Ocean with very inexperienced people at the controls, with instructions to follow the preceding plane. The lead place would be held by a plane with a crew that was able to fly and navigate by themselves. Incidentally, the pilot of the lead plane would be Paul Tibbets, known later as the pilot in the Hiroshima bombing. At the time of setting up the operation, the estimated loss rate was 10%. The effective rate was lower, although the operation would lead to a well-known crash-landing on the Greenland ice.

A few months later, some of the bombing raids during World War II involved more than a thousand planes and, of course, every one of them had its own crew. Very often, some non-pilot crewmembers would receive

FIGURE 6.1
Plane hanging from U.S. Navy air carrier. (Courtesy of Naval History and Heritage Command.)

basic pilot training since a single machine gun burst could kill both pilots. Furthermore, the number of people killed in action would lead to a frequent requirement for replacements and, of course, the replacement crews also had to be trained for the task.

Therefore, the massive use of Aviation introduced a new requirement: Aviation should be usable, at least for most of the tasks involved, by ordinary people whose skills, intelligence or fitness could be considered normal. Otherwise, manufacturing planes would be easier than finding people to fly them.

Some tasks would still require highly skilled people, such as flight tests, landing in unprepared places or performing special missions, like the first bombing missions over Japan, which departed from a carrier, something never seen before (Figure 6.2).

Therefore, Aviation became a mass activity that would require as ordinary people, while some specific tasks would require more specialized people. This situation still remains since, as already mentioned, the regulations repeat time and again that the handling of some systems must not require especially skilled or alert crews.

The requirement during and after World War II is easy to understand, but the necessity by itself did not give any clues about how that could be performed. The solution would be conceptually simple: an industrial approach – that is, mass production.

FIGURE 6.2
B25 taking off from a U.S. Navy air carrier. (Photo by Journalist 1st Class Joe Gawlowicz; courtesy of Naval History and Heritage Command.)

In some way, that is the rebirth of the *scientific organization of work* pioneered by Frederick W. Taylor (1914). For him,

> The managers assume, for instance, the burden of gathering together all of the traditional knowledge which in the past has been possessed by the workmen and then of classifying, tabulating, and reducing this knowledge to rules, laws, and formulae which are immensely helpful to the workmen in doing their daily work.

Probably, and with good reason, a reference to something written in 1911 can sound a little bit outdated, but before dismissing it we should remember that the much more modern *intellectual capital* wave was born under the worrying idea – from the shareholders' point of view – that the physical assets of many modern companies were far below the actual value. In the words of Edvinsson and Malone, "Human capital cannot be owned by the company." The real value of some companies was in the heads of their workers and in the social networks leading to get the best people and the best use of knowledge contents. In other words, the shareholders of companies such as Google, Facebook, Microsoft, Apple and many others felt that the value of their investments depended of the good will of others.

This situation, for intellectual capital prophets, was something to be managed and corrected – exactly the principles defended in 1911, written with other words. Of course, the resources available in the last few decades of past century have allowed for solutions other than mass production, but the main principle remains.

This principle has been present for a very long time: Anything that can be automated or performed through blind procedures – *know-how* as a replacement for *know-why* – can be performed by fewer people, who are less skilled and less trained and, hence, more easily expendable.

An environment based on individual skills and individual knowledge does not allow for this solution and, hence, it is something to avoid or to be allowed only in exception. In the middle of a war, many people were required, and things had to be built and managed in an "idiot-proof" fashion, which justified an industrial approach. However, it remained well after the war was over and, for different reasons, it still remains and has adapted to the new technological environment.

Technological evolution kept a very fast pace and the war between exceptional versus procedural individuals was highly visible during the first space missions and the pre-eminence of the astronauts over the best pilots of Edwards Air Force base. Actually, Charles Yeager, the first man to fly faster than sound, explained that space ships did not need pilots and the first missions would be performed by an automated capsule with a monkey as the only occupant.

It would be hard to find a single field where technology did not improve the qualities of performance and handling, always aiming at the same goal: Fewer and less-trained people should be required to handle the system.

Some functions that, formerly, were important tasks for flight engineers, such fuel and cabin pressure management, became fully automated. Navigation is no longer an art, and getting lost is hard, even for anyone unable to find the polar star or the cardinal points. Landing without visibility is more about programming and supervising systems than about piloting skills and so on.

In the meantime, while the number of events was decreasing, some people were speaking about a new factor – that is, technological understanding, especially when automation and Information Technology were involved.

> Accidents like AA965 can be easily explained today as a "smartphone mistake": To call someone, the first letter of the name can be dialed. Unfortunately, if there is another contact sharing the same first letter and whose second letter in alphabetical order comes before the intended receiver, someone will receive a misdial.
>
> By converting the contacts list of a smartphone (even an elementary one) into a navigation database with names and frequencies, the same thing can happen. Dialing "R" for "ROZO" in the database will lead to the wrong place if there is another nearby point called "ROMEO". If there is a mountain between ROMEO and the position of the plane, the identical action can result in a misdial or an airliner crash.

Of course, this was not the only factor in the AA965 case. The explanation is simplistic for the sake of clarity, but it can give the idea that, perhaps, converting paper charts into a navigation database could lead to unintended consequences that nobody was able to foresee. Many other cases can be mentioned where there seems to be a huge disproportion between the magnitude of the mistake and the magnitude of the consequences. For example:

- The Mount St. Odile case, where the same dial could be used to set the angle or regime of descent.
- AeroPeru 603, where a piece of tape in the static port was enough to induce full confusion in the crew.
- Air Transat and the loss of situation awareness linked to the automation of fuel management.
- Air Canada 143 and the fuel calculations.
- Emirates 407 having the wrong weight for the take-off power.
- Asiana and the erroneous interpretation of what the plane was doing.

All the aforementioned cases have something in common: They are "new"; that is, accidents related to mechanic faults, maintenance or errors in the operation kept happening, but there is a new type of accident, coming from confusion and from efficiency.

The defining point of efficiency is the difference between input and output. If a tiny input can lead to a huge outcome, we can say that is an efficient system and, as such, this system will also be efficient in its faults.

It must be noted that in the AA965, AC143 (Gimli glider) and AeroPeru examples, the planes were from the former generation (Boeing-757/767). Therefore, confusion-inducing technology is not a new phenomenon, but instead of it being solved, it kept growing.

Probably the most notorious event of this new type is the aforementioned AF447, where the pilot did not receive kinesthetic information from the flight controls, where one of the pilots could not see in his own controls what the other was doing, where the stall warning remained silent when the situation worsened and went off when something was being corrected, inviting the pilot to pull up (i.e. exactly what the pilot was doing wrong) – and all of this coming from a single failed sensor. In a confusing environment, two engines working perfectly were not enough to prevent the plane from crashing.

In the subsequent investigation, lack of training was considered the main factor leading to the accident, and in some ways the conclusion was right, but it was not necessarily legitimate.

Lack of training is not a valid explanation for an event since there is a structural lack of training coming from the management model: a model that has been with us at least since 1911, in which complex activities are performed by ordinary people, instead of relying on extraordinary and highly trained people.

That's why the conclusion can be, at the same time, correct and dubiously legitimate: Lack of training is not an accident but a piece of a system, whose evolution is based on it.

Complexity has been driven further and further out of common tasks, and embedded in technological and organizational processes. The critical part in the new fashion is the interface; that is, the system must appear as agreeable and as easy to handle as possible, while it hides the real complexity of the process inside.

Furthermore, in Information Technology lingo, they describe this situation as being *transparent* to the user. In ordinary language, being transparent means something is so clear that we can see through it. In Information Technology, transparency is a polite way to tell the user, "Mind your own business."

The user, in this model, should be centered on managing a nicely designed interface without worrying about what is inside. In other words, the user should be limited to something that is here called *Windows knowledge* – that is, the metaphor-based knowledge of a common Windows user about how computers work.

This state of things can be criticized but it is present virtually everywhere, not only in Aviation. Should this practice be criticized everywhere? Is there a good reason to criticize it precisely in Aviation?

Perhaps, from a social point of view, converting people into easily expendable assets should be criticized everywhere because this practice steals the meaning, and hence the motivation, for any task: It converts rational people into robots, whose behavior often becomes almost as lifeless and mechanical as that of an automatic device.

However, regarding Aviation, there is good reason to go beyond that point. Aviation has an uncommon organizational anomaly: The chain of command is physically broken, and any serious event must be solved at once – there is not a "pause" button available – and with the resources at hand on board.

That means that the old division between designers and performers, coming from Taylor's time, does not work here. Again, *acceptability* becomes the keyword. It would be interesting to test the acceptability of a statement like this:

> Aviation is exceptionally safe. That means that your chances of being killed in an accident are far below your chances of winning a big prize in the lottery. However, you should know that a very improbable situation could happen where, without a major technical failure, the pilot could get so confused that the plane crashes.

This statement is trying to capture the essence of a situation that, despite being real, will never be communicated in these terms to users.

Beyond statistical sorcery, many people would probably not accept confusion as a legitimate origin for a disaster. As Fischhoff et al. (1983) say,

"However mathematical their format, approaches to acceptable risk are about people; for an approach to aid the decision-making process, it must make assumptions about the behavior and, in particular, the knowledge of experts, lay people, and decision makers. When these assumptions are unrecognized or in error, they can lead to bad decisions and distortions of the political process."

In more detailed words, the *disaster threshold* concept can be found in this statement. That's why the *bad apple theory* – coined by Sydney Dekker (2017) – is so frequently used as an explanation for major events. Calling it "bad design" would be indelicate, especially if it referred not to the physical features of the plane but to the social design inspired by the old Taylor model.

A Comet-type unforeseeable technical disaster might be better accepted than a disaster coming from a design aimed at using cheap and easy-to-replace people – that is, an accident inspired by Taylor's mass-production principles: While the first case would be unforeseeable, the second would come from an assumed risk. The public answer would be very different.

As previously mentioned, an operator can decide between a "cheap license" or a "good pilot", but nobody should expect the same outcome in both cases when things become difficult. Competency and knowledge of procedures are not synonymous; the first is much costlier in both time and money and, additionally, people must be chosen very carefully.

Teaching people to perform procedures is a shortcut that, usually, works, but if something unexpected happens an adequate answer should not be expected. The organizational model has evolved in a way that, beyond limiting the competency that can be acquired through experience, has limited the options to make the right assessment, and hence the right intervention, in abnormal conditions.

Let's say, using the word *transparent* in its original sense – not that used in I.T. environments – that the organizational system composed of procedures, people and technology has become less and less transparent. Hence, obtaining knowledge from direct experience filtered by procedures and technology far from the "real thing" is not enough to reach the level of *knowledge*, in the sense used by Rasmussen – that is, know-why.

Therefore, although downgrading competency to procedural knowledge is always a temptation, when a major event happens, claiming "lack of training" is already a worn-out silver bullet.

Lack of training has become structural, as in one of the key pieces of structural engineering, not as a hole to be fixed in a system that allowed an untrained individual to be in the wrong place at the wrong time.

So, if a loss of acceptability is to be confronted, whatever the statistical records say about safety, the next question should be: What is the correct training?

What the New Training Principles Should Be

"Lack of training", once again, cannot be accepted as a final cause, since, if so, it points to an organizational issue:

Why was a person without the right training in the wrong place?

So, it is not an individual problem but organizational, but which organization? The company who employs the poorly trained person or the regulator who establishes what the correct training should be?

Very probably, the answer is the employer, because regulators flee from this disturbing situation: The regulator, regarding training, indicates general knowledge, sometimes very specific, but at the same time it can add some final clause. These final clauses should be read as fire escapes for the regulator because, based on them, the responsibility will point to the employer.

As an example, the word *controllability* appears 1,126 times in "EASA CS25 Amendment 22: Certification Specifications and Acceptable Means of Compliance for Large Aeroplanes", in reference to different conditions. So, if control of the plane is lost while meeting these conditions, "lack of training" would be claimed.

Again, that would be partially true. For instance, if a plane is designed in a way that stalling is not possible – and the regulator buys this from the manufacturer – pilots would not be trained to manage the situation. When an exceptional weather-related situation shows that, after all, stalling is possible, pilots will not be adequately trained to manage it.

Beyond this obvious problem, there is something more: Nobody can guarantee to provide the right training in a complex and *tightly coupled* organization (using Charles Perrow's terms).

The potential number and types of unexpected interactions is high enough to make the foresight of everything an impossible mission. Hence, the most common behavior is the inclusion of "patches" in the training programs as specific answers to unforeseen events that, afterwards, were stamped "lack of training".

Usually, manufacturers are asked to include in their designs emergency resources for system malfunctions, but they are harder and harder to use, especially in new and highly automated planes.

The problem with using these emergency resources is twofold: First, knowing clearly when they should be used. Second: Emergency resources can share some systems or sensors with common resources. So, if one of these shared resources fails, the emergency resource will not work, increasing the confusion of the situation.

- Going back to AeroPeru 603, when the pilots observed that the information they were receiving was absurd, they tried to use the most obvious "escape door" – that is, forgetting the screens of the cockpit, instead using the basic instruments.

  The solution did not work because both the advanced and basic instruments were taking the information from the same input and, in this case, the problem was not in the processing but precisely in the input: A piece of tape was blocking the sensor (the static port), leading to false information about both speed and altitude.

  The situation became still worse when they tried to use another resource that, again, had a shared input with the malfunctioning part of the system: the secondary radar.

  By trying to get guidance from ATC, a single fact was missed: The data about speed, course and altitude available to ATC came from the plane, which was supplying the wrong information. Hence, the guidance was not reliable at all.

- The XL888T case shows how difficult it can be to find the right emergency resource in some situations. It was blamed on lack of training for a flight that was recognized as a "non-test flight", and pilots were accused of trying a maneuver at a height where recovery could be difficult.

  Perhaps that is correct, but the whole point became spoiled when, a little bit later, the AF447 would show that, even at cruise altitude, confusion can lead to crash: In XL888T, when the plane began a steep and uncommanded ascension, the pilots tried to correct it, but the

FBW system was working on its own, with automatic actions coming from incorrect inputs received from two frozen sensors.

The investigation would reveal a solution to the puzzle: Using the trim, since it could be handled outside the FBW system, would allow the plane to be controlled. So, it seems that the crew had chosen the wrong way to solve the situation. The right way was harder to find on the go, but its existence was enough to blame lack of training.

Usually, the most common emergency resource is flying manually, while using different sets of instruments or systems or different inputs. That's fine but, very often, the design of the plane causes it to behave like a magician – that is, to create a distraction while the real action is in a different place.

For instance, receiving a bizarre set of data while the plane is flying normally will invite the crew to qualify the data as bizarre, as happened on QF32, where the systems of the plane were referring to a failure in engine 4 – that is, at the other extreme of the plane.

However, if the wrong set of data triggers an automatic action, the problem to be solved is the wrong automatic action and how to correct it: The automatic action distracts the pilot from the original problem, making it harder to assess and fix.

The tightly coupled model of Perrow works both in organizational design and in physical design. A faulty sensor can trigger a set of false warnings, an automatic action or both. So, in a very short time, a pilot is supposed to be able to

1. Identify the information as potentially false
2. Be aware of the danger linked to the possibility that the information could be, after all, correct
3. Figure out the process that has led to the false information and which systems could be affected
4. Overrule an automatic process that is running because of the information

It seems too much for a limited time span. The common solution used by manufacturers and accepted by regulators is to use up to three systems working in parallel. In that fashion, the discrepant system can be identified as faulty and ignored, a practice that someone brought to the cinema as *Minority Report*.

Unfortunately, as in the movie, some situations can appear where the correct system is the discrepant one, as the XL888T case showed.

As previously mentioned, the organizational model used to decide what kind of people and what kind of training should be used can be traced to 1911: ordinary people with the adequate knowledge for common operations. This model does not work in Aviation, due to an organizational feature where serious issues must be solved locally and at once. So, before going into detail, the features that a good model should have will be analyzed.

A relevant description can be found in the work of Daniel Dennett (2008) and his division of organizational models in a framework with two extreme types: Information Agency and Commando Team.

The Information Agency works under the *need-to-know* principle – curiously, a concept used by one of the main Aviation manufacturers to define training requirements – and the idea that everyone must receive only the information required for the specific task.

Since they try to avoid major leaks of information, this principle helps to keep control over the information flow. In a different way, major industrial companies are also afraid of the possibility of having their secrets revealed to competitors, and giving fragmented views or metaphorical approaches – what is called Windows knowledge – can help, but its side-effects have appeared in different events.

The opposite model to the Information Agency would be the Commando Team. In the words of Dennett, the Commando Team principle means "giving each agent as much knowledge about the total project as possible, so that the team has a chance of ad-libbing appropriately when unanticipated obstacles arise". That's because a Commando Team is sent on difficult missions where different contingencies and casualties should be expected.

That means a different recruitment process where special people are required. They can be physically exceptional, especially skilled in some field or both. Additionally, there must be a partial overlap in the skills to make sure that losing a member does not make meeting the objectives impossible.

Unlike the Information Agency, the problem with the Commando Team is not potential information leaks but the potential failure of the mission. That's why there are redundant resources – to keep the mission feasible, whatever the contingencies could be.

Which model is better? The Commando Team model implies an expensive recruitment process, looking for uncommon skills, and expensive training, explicitly looking for redundancy and a surplus of knowledge.

Of course, the most expensive part is the expected shortage. If a recruitment process is aiming at finding special people, those special people are, by definition, scarce. Is it affordable?

On the other hand, the Information Agency model is cheaper and has a lower risk of information leaks, but if something unexpected happens, a sound answer should not be expected.

Does this framework apply to Aviation and, if so, what are the problems? Some specific tasks still require the Commando Team model – flight tests and some military-related activities would be good examples – but, for the main part of the activities, the Information Agency model would be chosen. So, what if something unexpected happens?

The answer to that is quite straightforward: technology, especially Information Technology. Right now, the capacity to store and process information is so great that the humble mobile phone everyone has in their pocket has more processing power than a computer manufactured only a few years ago, not to say a computer like the one on board *Apollo 11*.

Apparently, unexpected situations should be planned out. Even so, things are not so easy.

- There is an increasing efficiency requirement in every imaginable field: More planes are packed in less airspace and many tasks have become automated, allowing them to be performed by fewer people or, simply, eliminated altogether.
- Training time is shortened through commonality and automation; crowded communication channels are converted into data links to increase bandwidth; fuel-efficient but unstable planes are flown through automated protections, making them suitable for non-exceptionally skilled pilots; and automatic landings can be performed with poor visibility.

That leads to an internal complexity and a potential snowball of events that could otherwise be easily dismissed or managed. If a frozen sensor, a piece of tape or moving a switch the wrong way can lead to a major event, something is wrong in the design of the whole system.

Surprisingly, identifying what is wrong is not difficult at all: Unlike chess, Go or other fields where computers can excel because they are in a closed environment, unexpected situations cannot be ruled out in Aviation. However, the system, as designed, shows a seriously limited ability to manage them. At the same time, the process to maintain the effectiveness of the human side, once the situation is beyond the control of the system, has failed.

It's not only a matter of training, following the Commando Team principles. It's also a matter of keeping people in the loop, making the system understandable – well beyond the physical ergonomics of the interface – and ensuring there is an early warning, to be fully aware of the sequence of events.

Air safety has improved over time. That should not be denied, but looking at new events – not only in the sense of recency, but in the sense of triggering situations unthinkable some years ago – an uncomfortable conclusion could be reached:

Foreseen events are managed better than ever and, hence, events related to them decrease. At the same time, new and unforeseeable events appear, and the system is not well equipped to deal with them, especially because many of them are not external, but coming from the system itself.

Since the number of potential foreseeable events is higher, safety rates have improved, but there is still a transaction: Better at doing what is already known and worse at doing what is unknown. Is that transaction acceptable?

From a statistical point of view, the answer should be a matter of numbers: If the number of prevented known situations is higher than the bad outcomes coming from unknown situations, it pays.

From a social and organizational point of view, the answer is not so clear: People will not accept a situation where a plane can fly but the wrong information, together with the wrong automatic processes, can produce a crash, while pilots are powerless to avoid it.

Tesla's irritating answer to the first fatal accident with autopilot showed the difference between both points of view – Telling someone that a relative has died, but that statistical reports show that the chances of being killed is higher in a human-controlled situation, is not convincing.

Confronting something that has already happened with calculated chances, showing that people have died because of an uncommon star conjunction, does not work. The disaster threshold concept is lost on the way, despite being important to understanding how to deal with acceptability.

Perhaps there is a way to escape the dilemma, and that will be the objective of the next chapter.

Bibliography

Australian Transport Safety Bureau (2011). Investigation number AO-2009-012: Tailstrike and runway overrun: Airbus A340-541, A6-ERG, Melbourne Airport, Victoria, 20 March 2009.

Australian Transport Safety Bureau (2013). AO-2010-089 Final investigation: In-flight uncontained engine failure, Airbus A380-842, VH-OQA overhead Batam Island, Indonesia, 4 November 2010.

Bainbridge, L. (1983). Ironies of automation. In: *Analysis, Design and Evaluation of Man–Machine Systems 1982* (pp. 129–135). Oxford: Pergamon.

Beck, U. (2002). *La sociedad del riesgo global.* Madrid: Siglo XXI.

Bennett, K. B., & Flach, J. M. (2011). *Display and Interface Design: Subtle Science, Exact Art.* Boca Raton, FL: CRC Press.

Buckton, H. (2016). *Friendly Invasion: Memories of Operation Bolero 1942–1945.* Gloucestershire, UK: The History Press.

Bureau d'Enquêtes et d'Analyses pour la sécurité de l'aviation civile (1993). Official report into the accident on 20 January 1992 near Mont Sainte-Odile (Bas-Rhin) of the Airbus A320 registered F-GGED operated by Air Inter.

Bureau d'Enquêtes et d'Analyses pour la sécurité de l'aviation civile (2010). Report on the accident on 27 November 2008 off the coast of Canet-Plage (66) to the Airbus A320-232 registered D-AXLA operated by XL Airways German.

Bureau d'Enquêtes et d'Analyses pour la sécurité de l'aviation civile (2012). Final report on the accident on 1st June 2009 to the Airbus A330-203 registered F-GZCP operated by Air France Flight AF 447 Rio de Janeiro–Paris.

Business Insider (Apr. 2, 2018). If you have a tesla and use autopilot, please keep your hands on the steering wheel. https://www.businessinsider.com/tesla-autopilot-drivers-keep-hands-on-steering-wheel-2018-4?IR=T.

Dekker, S. (2017). *The Field Guide to Understanding 'Human Error'.* Boca Raton, FL: CRC Press.

Dennett, D. C. (2008). *Kinds of Minds: Toward an Understanding of Consciousness.* New York: Basic Books.

Dismukes, R. K., Berman, B. A., & Loukopoulos, L. (2007). *The Limits of Expertise: Rethinking Pilot Error and the Causes of Airline Accidents.* Aldershot, UK: Ashgate.

Dismukes, R. K., Kochan, J. A., & Goldsmith, T. E. (2018). Flight crew errors in challenging and stressful situations. *Aviation Psychology and Applied Human Factors,* 8(1), 35–46.

EASA, CS25 (2018). Certification specifications for large aeroplanes: Amendment 22. https://www.easa.europa.eu/certification-specifications/cs-25-large-aeroplanes.

Edvinsson, L., & Malone, M. S. (1997). *Intellectual Capital: Realizing Your Company's True Value by Finding Its Hidden Brainpower.* New York: Harper Business.

FAA Federal Aviation Administration (2016). Human Factors design standard. FAA Human Factors Branch.

Fischhoff, B., Lichtenstein, S., Slovic, P., Derby, S. L., & Keeney, R. (1983). *Acceptable Risk.* Cambridge: Cambridge University Press.

Gabinete de Prevençao e Investigaçao de Acidentes com Aeronaves (2001). Accident investigation final report: All engines-out landing due to fuel exhaustion Air Transat Airbus A330-243 Marks C-GITS Lajes, Azores, Portugal, 24 August 2001.

Hale, A. R., Wilpert, B., & Freitag, M. (Eds.) (1997). *After the Event: From Accident to Organisational Learning.* New York: Elsevier.

Hubbard, D. W. (2009). *The Failure of Risk Management: Why It's Broken and How to Fix It.* New York: John Wiley

Johnson, W. G. (1980). *MORT Safety Assurance Systems* (Vol. 4). Marcel Dekker.

Klein, G. (2013). *Seeing What Others Don't.* New York: Public Affairs.

Klein, G. A. (1993). *A Recognition-Primed Decision (RPD) Model of Rapid Decision Making* (pp. 138–147). New York: Ablex.

Ladkin, P. (1996). AA965 Cali accident report near Buga, Colombia, December 20, 1995. Prepared for the WWW by Peter Ladkin Universität Bielefeld Germany. http://sunnyday.mit.edu/accidents/calirep.html.

Leveson, N. (2011). *Engineering a Safer World: Systems Thinking Applied to Safety.* Cambridge, MA: MIT Press.

Luhmann, N. (1993). *Risk: A Sociological Theory.* New York: A. de Gruyter.

Maurino, D. E., Reason, J., Johnston, N., & Lee, R. B. (1995). *Beyond Aviation Human Factors: Safety in High Technology Systems.* Aldershot, UK: Ashgate.

Ministry of Transport and Civil Aviation (1955). Report of the court of inquiry into the accidents to Comet G-ALYP on 10th January 1954 and Comet G-ALYY on 8th April 1954.

Ministry of Transports and Communications-Accidents Investigation Board (1996). Accident of the Boeing 757-200 aircraft operated by Empresa de Transporte Aéreo del Perú S.A. 2 October 1996. https://skybrary.aero/bookshelf/books/1719.pdf.

National Transportation Safety Board (2014). NTSB/AAR-14/01 Descent below visual glidepath and impact with seawall Asiana Airlines Flight 214 Boeing 777-200ER, HL7742 San Francisco, California, July 6, 2013.

Perrow, C. (2011). *Normal Accidents: Living with High Risk Technologies-Updated Edition.* Princeton, NJ: Princeton University Press.

Raskin, J. (2000). *The Humane Interface: New Directions for Designing Interactive Systems.* Reading, MA: Addison-Wesley Professional.

Rasmussen, J. (1986). *Information Processing and Human–Machine Interaction: An Approach to Cognitive Engineering, North-Holland Series in System Science and Engineering, 12*. New York: North-Holland.

Rasmussen, J., & Vicente, K. J. (1989). Coping with human errors through system design: Implications for ecological interface design. *International Journal of Man-Machine Studies*, 31(5), 517–534.

Reason, J. (1990). *Human Error*. Cambridge: Cambridge University Press.

Reason, J. (1997). *Managing the Risks of Organizational Accidents*. Aldershot, UK: Ashgate.

Reason, J. (2017). *The Human Contribution: Unsafe Acts, Accidents and Heroic Recoveries*. Boca Raton, FL: CRC Press.

Rudolph, J., Hatakenaka, S., & Carroll, J. S. (2002). *Organizational Learning from Experience in High-Hazard Industries: Problem Investigation as Off-Line Reflective Practice*. MIT Sloan School of Management Working Paper 4359-02.

SAE (2003). Human Factor considerations in the design of multifunction display systems for civil aircraft. https://www.sae.org/publications/technical-papers/content/1999-01-5546/ARP5364.

Sagan, S. D. (1995). *The Limits of Safety: Organizations, Accidents, and Nuclear Weapons*. Princeton, NJ: Princeton University Press.

Saint-Exupéry, A. (2002). *Wartime Writings, 1939–1944*. Boston, MA: Mariner Books.

Saint Exupéry, A. D., & Cate, C. (1971). *Southern Mail and Night Flight*. New York: Penguin Modern Classics.

Sánchez-Alarcos Ballesteros, J. (2007). *Improving Air Safety through Organizational Learning: Consequences of a Technology-Led Model*. Aldershot, UK: Ashgate.

Senge, P. M. (2010). *The Fifth Discipline: The Art and Practice of the Learning Organization*. London: Cornerstone Digital.

Sowell, T. (1980). *Knowledge and Decisions* (Vol. 10). New York: Basic Books.

Stanton, N. A., Harris, D., Salmon, P. M., Demagalski, J., Marshall, A., Waldmann, T., et al. (2010). Predicting design-induced error in the cockpit. *Journal of Aeronautics, Astronautics and Aviation*, 42(1), 1–10.

Taylor, F. W. (1914). *The Principles of Scientific Management*. New York: Harper.

Tesla (2018). Full self-driving hardware on all cars. https://www.tesla.com/autopilot.

Transportation Safety Board Canada (1985). Final report of the board of inquiry investigating the circumstances of an accident involving the Air Canada Boeing 767 aircraft C-GAUN that effected an emergency landing at Gimli, Manitoba on the 23rd day of July, 1983 Commissioner. The Honourable Mr. Justice George H. Lockwood, April 1985.

Vicente, K. J. (2002). Ecological interface design: Progress and challenges. *Human Factors*, 44(1), 62–78.

Vicente, K. J., & Rasmussen, J. (1992). Ecological interface design: Theoretical foundations. *IEEE Transactions on Systems, Man, and Cybernetics*, 22(4), 589–606.

Walters, J. M., Sumwalt, R. L., & Walters, J. (2000). *Aircraft Accident Analysis: Final Reports*. New York: McGraw-Hill.

Weick, K. E., & Sutcliffe, K. M. (2011). *Managing the Unexpected: Resilient Performance in an Age of Uncertainty* (Vol. 8). John Wiley.

Wheeler, P. H. (2007). *Aspects of Automation Mode Confusion*. Doctoral dissertation, MIT, Cambridge, MA.

White House Commission on Aviation Safety and Security (1997). Final report to President Clinton, February 12, 1997. https://fas.org/irp/threat/212fin~1.html.

Wiegmann, D. A., & Shappell, S. A. (2003). *A Human Error Approach to Aviation Accident Analysis: The Human Factors Analysis and Classification System*. Surrey, UK: Ashgate.
Winograd, T., Flores, F., & Flores, F. F. (1986). *Understanding Computers and Cognition: A New Foundation for Design*. New York: Addison-Wesley.
Wolfe, T. (1979). *The Right Stuff*. New York: Farrar, Straus and Giraux.
Wood, S. (2004). Flight crew reliance on automation. CAA Paper no. 10.

7

The Future of Improvements in Air Safety

Once we analyzed the situation, the obvious question should be "What next?"

The equally obvious answer, from a technical point of view, would be "Keep going"; that is, we should use all the potential of information technology (IT) to decrease the training of the involved people as well as the number of them; at the same time, we should increase efficiency everywhere.

The last attempt in that direction is the *single pilot* proposal advanced by the CEO of Ryanair, Michael O'Leary, but more recently it has been addressed by FAA and Boeing. So, it seems that the market is preparing for that next step, after a long process of devaluating the first officer role.

Of course, the single pilot scenario would require more sophisticated systems than those already present in any modern plane, and some people are talking about the next big thing – that is, Artificial Intelligence (AI) and Machine Learning. In other words, the system will advance on the same path – no changes.

A milestone for Machine Learning has already been set by the AlphaZero program. AlphaZero was designed to learn and, after being given the rules of chess, the system was left on its own for 4 hours (an eternity in terms of processing time, even though it is very short for the human scale) playing against itself and trying options.

The acquired learning was enough to defeat the most advanced chess engine at the time, Stockfish, which despite its power could be considered humanlike, in the sense that it stores specific knowledge supplied by humans.

This event should not be overlooked by anyone trying to avoid essentialist positions about the roles of humans and technology. However, it must be remembered that chess, Go or any other complex game are environments with fixed rules, whose number of combinations are virtually infinite – at least, well beyond human reach. Hence, this is the perfect environment for advanced IT.

What if the environment makes rules unusable or rules change on the go? Nobody needs to invent situations because someone has already done it: Isaac Asimov (1951). His Three Laws of Robotics seemed to guarantee full safety for humans.

1. A robot may not injure a human being or, through inaction, allow a human being to come to harm.
2. A robot must obey the orders given it by human beings except where such orders would conflict with the First Law.

3. A robot must protect its own existence as long as such protection does not conflict with the First or Second Laws.

Even someone so unsuspicious of AI as Rodney Brooks (2003) would recognize that the problem, right now, is that we don't have robots intelligent enough to follow these rules. Anyway, let's ignore that limitation. Let's suppose that this is possible.

Despite these apparently unbreakable rules, many of Asimov's books are devoted to situations where these rules are not enough to protect humans.

Robots, in the novels by Asimov, are supposed to be faster and smarter than humans and without bugs in their processes beyond those related to the three basic laws.

From the science fiction point of view, a perfect algorithm without software bugs is compulsory, but in real life minor failures in algorithms – not only in the major laws – are a common fact. Asimov himself introduced the idea of software bugs in one of his books, *Azazel*, where a very powerful devil, but without knowledge of the environment, was keen to please his master, forgetting some key details in the process.

That's a very common source of problems in real life. Beyond big principles in the design, little mistakes in the millions of program lines can lead to unexpected situations.

A very well-known case that became public was the issue with the Pentium processor designed by Intel and running the vast majority of personal computers of the time. We might see this as being far removed from the strict safety controls of Aviation.

If so, it would be worth remembering that the Boeing 777, still leaving the Boeing factory at a good pace, has an Intel 80486 processor – that is, precisely the former generation to the flawed Pentium.

Designers and regulators have always tried to keep a firescape in the system (though it is not always effective), and this firescape is based on redundancy, intended or unintended, allowing the user to manage the system in degraded modes. The unintended part will be the most relevant from Human Factors point of view.

When automation and screens began to appear in planes, basic instruments were compulsory. The idea behind this requirement was clear, and could be explained in these terms:

> Everything is fine, but we don't fully trust the new systems. If something fails, you must be able to go back to a former development stage and solve the problem the old way.

An experienced old-school pilot rephrased this principle in a funny way: "Give me a plane as automated as you want, as long as it has a big yellow button that converts it into a DC9."

However, in some cases we should ask if this is enough.

- Gimli Glider: A plane ran out of fuel and tried to land on a short runway. The pilot had to lose height without increasing speed and the old school had a solution for that, albeit unusual in a commercial environment: sideslip.

  As well as the basic rules, would an eventual AlphaZero program learn this option? Would it try by itself in a trial-and-error fashion during the machine-learning process? Would the system try to land there knowing that the performance data were against it?
- QF32: The uncontained explosion of an engine severed different lines, including data lines. The pilots, after concluding that they were receiving false information, decided to ignore the data from the plane, trying to establish valid information using their own resources.

  What kind of resources? For instance, if the pilot did not trust the information about the stall speed, the plane could enter a near-stall situation to see when the first symptoms appear. Again, would the system be able to overcome a situation of corrupt data and try other resources?
- XL888T: A kind of "voting" performed by three systems led them to ignore the one that was offering discrepant information.

  Unfortunately, the discrepant system was correct, and the automatic system followed the indications of the other two. Would an advanced system be able to check this?

These and some other situations can be taught as a part of the basic rules of the system. Additionally, it can be tested to see if the Machine Learning process has reached that point in a process of *unsupervised learning*.

However, the real problem is that, in an open environment, it is only possible to include unexpected situations *once they have happened* because, before that, no one is able to foresee the situation. In other words, they can only be managed once they are no longer unexpected. The only option for them is thinking out of the box.

Someone in the "strong AI side" could say that many more situations could be foreseen in an unsupervised learning process. That is probably right, but it could lead to a different and more serious danger.

Unrealistic scenarios could be developed together with realistic ones without clear criteria to distinguish between them or even to detect their existence before an event exposes them. At the moment of decision, the system would not have – as humans do – external criteria to decide. If a system designed for shape recognition can confuse a helicopter with a rifle,[1] it seems clear that it could react by using an unrealistic option based on that perception. So, unrealistic potential scenarios are not free; they introduce noise that can lead to poor evaluations.

Once it has happened, it is possible to introduce criteria to decide whether uncommon data are coming from a failed sensor (AF447 or XL888T) or look for alternatives if the flight controls become unusable[2] (U232), but many new situations can still arise where the basic rules cannot guarantee that the learning process will come up with the right solution.

This is one of the commonly disregarded dangers of unsupervised learning: A system trying to learn will probably process more information than a person will in their whole lifetime. No one can process the same amount of information from the internet at the same speed as a connected system. The strong points for a human being are the possibility of jumping to a conclusion from far fewer data than a computer and having an exclusive source of information – that is, the information coming from their body and all its specific sensors.

Obviously, there are two very different models working in a very different fashion: a frugal model (human) and an intensive data model (an unsupervised learning system). In both cases, the quality of data input should be something to be worried about – that is, being sure that the required relevant data are present and there are no data that can misguide the process. The person has background knowledge that can be used as a filter, while the system can only "vote" – that is, assume the information that appears more frequently is certain.

Even without mistaken content, so frequent in sources like the internet, contradictory information can come from different perspectives. Although there is no mistake, different interpretations of the same facts can lead to contradictions and they are difficult for a system to manage.

Humans are also subject to these contradictions, but there is a way to manage them other than using a kind of poll that would be, at least, confusing:

People choose the perspective from which to analyze the problem and, once chosen, it is used as a guide to assess the validity of the incoming data. Data that don't fit with this perspective can be rejected without processing them as a contradiction to be managed. They are simply ignored, and no further processing resources are wasted. These incoming data can be ignored because they are considered erroneous or because they come from an approach that is not relevant to the specific situation.

Sometimes, data can come from a different perspective but, despite that, they can offer a solution to a specific problem. They could be called a fertile fallacy – being used despite knowing that they don't fit with the wide perspective used to manage the situation. These data can be "isolated" in a kind of microworld, which is useful for solving specific problems but without an overall perspective.

In other words, people are equipped to deal with contradictory data: First, a perspective is chosen; second, the contradictory data can be rejected or accepted but isolated. That is not a processing problem for people, but it could be extremely hard for a system trying to learn in an unsupervised fashion.

If filtered in advance, the sources of data still cannot guarantee all the required knowledge and, if not filtered, the processing of contradictory data – because of being erroneous or because of coming from different perspectives – cannot guarantee that the system will not reach bizarre conclusions, applying them at the worst moment, through the wrong assessment to the human operator or, worse still, through the wrong direct action.

The previous examples suggested as potential tests for AI or Machine Learning do not mean dismissing the potential of these tools. Simply, there are serious warnings about their use in high-risk environments.

It's true that we can identify some situations in the past that could have been solved in a very short time by using big data – had it existed at the time, of course. The discoveries of the real natures of sicknesses like cholera or AIDS were based on handling huge amounts of data and coming to amazing conclusions from them.

Now, what can be expected of these systems in a time-constrained environment where, perhaps, the behavior of the system is challenging the known principles? One of the features of human beings is their ability to change focus, looking for relevant information in unexpected places. Will the new systems be able to do that?

Defenders of an IT-based mainstream would have two different arguments:

1. Statistics: Statistical reports say that advanced automation can be safer than humans. Failures can happen, but we should have fewer accidents than when using human operators. This approach, related to the fatal accident in a Tesla car driven by the autopilot, can be found in the statement by the company that designed the car:

   > Tesla declined to comment on the California crash or to make Mr. Musk or another executive available for an interview. In its blog post on Friday about the crash, the company acknowledged that Autopilot "does not prevent all accidents," but said the system "makes them much less likely to occur" and "unequivocally makes the world safer".

2. Knowledge of human operators: The algorithmic approach can be criticized, but it is impossible to know whether, in unforeseen situations, human knowledge will be adequate to manage the event.

Perhaps a situation could be reached where automation leads to fewer accidents than human operation. Even so, the acceptability issue, which goes beyond statistics, should be managed. Users will not accept a situation where a last resource is missing, based only on the favorable chances of managing the event using the resources designed for common operation.

Going back to the root of the discussion, Turing's ideas on intelligent machines and the discussion that came with it seem relevant:

For Turing, a machine could be considered intelligent if that machine was able to mislead a person interacting with it, leading this person to think that he or she was interacting with another person, not a machine.

John Searle replied with the so-called Chinese room experiment: Two people are speaking through cards written in Chinese. One of them is a Chinese speaker and is looking for the right card to answer. The other does not speak Chinese, but there is a list, indicating which is the right card to use in the interaction. At the end of the interaction, the Chinese speaker thinks that the other person is also a Chinese speaker. Searle asks whether this belief qualifies someone as a Chinese speaker or not. The answer is clear: It does not.

Turing's initial idea and Searle's reply have been revisited at least twice, with some elements added that could clarify and be applied to the present situation:

- Rodney Brooks, former director of the MIT Computer Science and AI Laboratory, commented that even though the person did not understand Chinese, the system, as a whole, did. The statement is quite clarifying because it clearly shows a confusion between two terms that are far from being synonyms: *understanding* and *operating*.

  The machine can operate, and if sensors, actuators and algorithms are properly designed, it can drive someone to take operational skills for understanding, meeting Turing's conditions for an intelligent machine.

  Anyway, things don't finish there: Even if dealing with both concepts – understanding and operating – as synonyms is not acceptable, this does not answer a previous question: Is understanding required if operational skills can perform the task?
- Jeff Hawkins's ideas about the Turing–Searle controversy could help to answer that question. For Hawkins, the real difference comes in a single question: What's coming next? If it is possible to foresee what is in the next scenario, someone/something – human or machine – could be said to understand a situation, whether it is a Chinese conversation or any other.

  In the words of Hawkins, "Understanding cannot be measured by external behavior ... it is instead an internal metric of how the brain remember things and uses its memories to make predictions."

  However, a contradiction can be observed: If, through understanding, someone can foresee what is coming next, understanding cannot be downgraded to a kind of internal state or *internal metric*. Understanding has its roots in the external world, and any forecast is about interacting with that external world. It's not simply an internal state.

That brings back the *situation awareness* concept, as defined by Endsley, where deep understanding can be seen as a feature of the third situation awareness level (Figure 7.1).

Closed environments with a high number of options – such as chess or Go – can be successfully operated by knowing and practicing the rules, while open environments, where new events can challenge the commonly used rules, require a high situation awareness level.

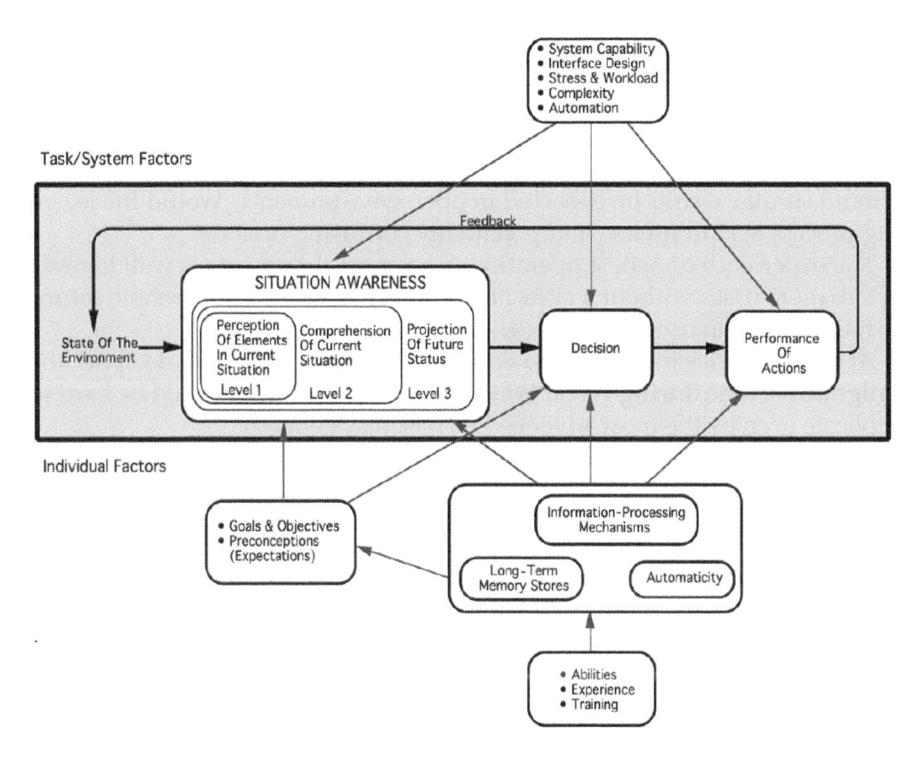

FIGURE 7.1
Endsley's model of situation awareness. (From Endsley, M., and Endsley, M. R., Toward a theory of situation awareness in dynamic systems. *Human Factors*, 37(1), 32–64, 1995.)

Furthermore, we should not forget that some new events come precisely from malfunctions that, once acknowledged by the operators, have some "hidden doors" to be managed. In other words, operators often manage these known malfunctions by hacking the system, to make it work as expected.

Of course, plain automation cannot reach that level: Automatic processes triggered by bad sensors (AF447 and XL888T) or autopilots carrying dead passengers after a loss of pressure (Payne Stewart accident) show that.

Many cases challenging for a Machine Learning process can be found, if there is no real consciousness about what is happening, and the potential performance is downgraded to a set of algorithms, whether they are fed to the machine (supervised learning) or the machine builds it by itself (unsupervised learning).

Cases like QF32, where systems are fed with corrupt data, or BA38, where the pilot lifted the flaps as the only available way to increase the gliding distance, can be converted into algorithms once they have happened, but not before.

Unforeseen situations, by definition, are not included in training plans, until an event shows that they should have been foreseen.

Meanwhile, the only available resource for these situations is precisely a high level of situation awareness and, nowadays, that brings the problem back to humans.

It is not a matter of underestimating Machine Learning and its potential. The AlphaZero case shows an impressive potential in closed environments. Should similar results be expected in open environments? Would the learning process lead to useless and potentially confusing options?

The experience of people operating in open environments is well known: A situation arises without a clear procedure to manage it and people return to basics or try to use valid analogies.

At this point, it's interesting to remember that the process to decipher the Enigma machine during World War II, led by Alan Turing, would be hard to replicate even by the most advanced AI based systems.

> In short, the Enigma machine had so many possible combinations that decrypting it was virtually impossible. Turing decreased the options to a number that a computer of that age could handle. To do so, he introduced two factors:

1. In some situations, he could be able to know at least a part of the contents of the message.
2. The original message was written in German.

Turing introduced new elements, coming from knowledge external to the situation. Of course, the brilliant part of the solution was not background knowledge that, probably, anyone in the same environment could share. The brilliant part was deciding which part of that background knowledge could be used and how to do it.

Going back to Machine Learning, it is dubious that a machine could acquire the kind of knowledge that could be considered irrelevant. Even if the machine is given access to the biggest data repository in the world – that is, the internet – it is still dubious that it would cross-reference supposedly irrelevant information (e.g. they spoke German) or make an inference such as "If the submarine is near to a convoy, the contents of radio emissions will probably speak about the position. Knowing that will help to decode other messages."

In some ways, the work performed by Turing is the real test for the differences between human and machine features, and is far more informative than his idea about an intelligent machines.

A good part of the process of managing a situation implies pruning non-valid options, and to do so, external items must be added to understand the situation and, with that acquired understanding, to manage it. Some data cannot be processed at a specific level but are easier to handle at others.

Seeing the problem from different angles and distances and introducing content and analogies from different fields is out of reach for the present state of IT, but it is a common human approach. In the case of Turing, an easy shortcut to these different angles would be something so common as trying to put himself in the shoes of a German submarine captain: "What kind of information would I transmit from this position?" and, of course, "What language would I be speaking?"

Of course, many human tasks don't require that approach. In closed environments, where rules are permanent, even if there is a virtually infinite number of options, this capacity could be traded for a fast and perfect performance, limited to that environment.

In open environments – that is, environments where new situations challenging existent knowledge can appear and where these situations can have a high impact – things should be designed in a way that this capacity to introduce new elements is guaranteed.

That capacity to change the focus and the angle to analyze a situation is linked for many people to consciousness. Strong AI supporters can be found who believe that consciousness is a useless feature and, on the opposite side, scientists such as Roger Penrose defend it as something out of reach for computers and with a clear value.

In some ways, consciousness behaves as a jazz band leader; that is, since there are contradictions among the stored data and incomplete information, the leader is the one deciding who is going to play the next solo and how the remaining members have to play to avoid disharmony at different moments.

AI could be more similar to an orchestra conductor – that is, someone with a full orchestra, where everyone has a music sheet and is playing at the right tempo and volume and the whole piece is written, and no surprises should be expected.

Every option has specific advantages, but from the acceptability point of view, it would be interesting to know how many people would feel comfortable being flown by a set of algorithms, more precise than a human pilot and, at the same time, unknowing of the single fact that they were flying, unknowing even what flying is and not especially concerned about their own survival.

Acceptability will always be an issue, and despite public relations efforts, any major event due to the wrong behavior by a system could lead to a sudden loss of that acceptability.

Luhmann (op. cit.) points out an unusual difference between the concepts of risk and danger. The definitions can be accepted or not, especially since they are not coincident with the most commonly used, but, accepted or not, there is something important in them:

> There are then two possibilities. The potential loss is either regarded as a consequence of the decision, that is to say, it is attributed to the decision. We then speak of risk – to be more exact, of the risk of decision. Or the possible loss is considered to have been caused externally, that is to say, it is attributed to the environment. In this case we speak of danger.

Let's say that this is not a common definition of danger. Usually, a danger is anything that has the potential to cause harm; and risk is the likelihood of danger arising, combined with the effect of that danger. However, despite the uncommon definition, there is a relevant single fact: People can accept danger – in the definition of Luhmann – in a kind of fatalistic way, while at the same time they cannot accept a major negative event if it comes from a decision, even – or especially – if that decision appears to be supported by statistics and technical knowledge without any other consideration.

That's the lost dimension in the evolution of the system: People can accept unforeseen harm, but those same people can reject the acceptance of a foreseen harm, based on a calculation about supposedly low chances.

So, there are strong reasons to keep people inside the system instead of trusting the life of passengers to systems that are able to perform programmed tasks but which become blocked or misbehave in unforeseen situations. Telling someone that statistics say these unforeseen situations are scarce is not acceptable, especially if a weird behavior is found in a situation that is easy to manage for humans.

However, a question still remains: What rules should be respected to get the best from both parts, human and systems?

The "PISS" Rule: Produce Intelligible Software, Stupid

A little thought experiment can demonstrate the main point: the librarian.

> There is a library with a lot of cards, ordered by title, author or subject, and a librarian is employed to handle them. Suddenly, someone comes up with the idea of removing the cards and, in their place, setting up a database.
>
> A programmer is called to work on the project and, after a short time, there is a design with simple linked tables and where the most common queries are included in a form that appears on the librarian's screen.
>
> So far, so good, but suddenly the system crashes. The old system has disappeared, and the librarian cannot work until the database is restored. What should be the solution to a similar future event?

There are, at least, three different solutions to the problem:

1. Assume this as a non-life-or-death problem and keep the database, with the support of the librarian or even in a self-service fashion, where the users could look for the books themselves.
2. Redundancy: The old system is kept together with the database. Of course, that means duplicating some tasks since, to be valid, the old system must be maintained.

3. Redesign: The system should be designed in a way that, if something fails, the librarian should have resources outside the system. These should be enough to perform basic tasks, even if the degraded mode could imply more time or the loss of some features.

In the example, the second option should be ruled out since the cost hardly compensates for the added functionality. In a critical-safety system, that would be a real option, unless the design of the system itself makes impossible it.

Let's suppose that the first option, commonly accepted in some low-risk environments, is also ruled out; that is, we won't accept that things won't work under some circumstances without the resources to replace the failed ones.

That would bring us to the third option, but, if so, how should a redesign be done?

Probably, the resulting design would be far less elegant than the design of the programmer and, ideally, prepared by the librarian. The librarian, if asked to design the system, could be expected to reproduce the old cards system to be used with a keyboard and a screen.

- The good part: When something fails, perhaps the librarian cannot access the forms, but the information would be available in tables whose formats were familiar to the librarian. Looking directly at the tables instead of the database, much of the library could be managed, even though the pace would be slower.
- The bad part: The solution is far less efficient than the solution designed by the programmer.
- The conclusion: Trading efficiency for intelligibility could be a sound business, especially in a field with high stakes like Aviation.

Different attempts have been made to avoid these problems – in particular, ecological interface design (EID). It should be noted that the idea of *interface design* suggests a change limited to formats, but the proposed change is far deeper:

It tries to keep the user aware of what is going on through the *abstraction scale* – that is, the idea of the user jumping at will from level to level of functional design, not only the interface. Of course, that means that the different levels must be understandable to the user. That would be a major change, compared with a situation where it's supposed – from the perspective of designers – that the best human contribution would be a blind procedural compliance.

If being understandable in different situations becomes a condition for an interface, it seems that the main effort would be on the software side. One could claim that introducing that condition could affect the software speed in a situation where there are clear time constraints. However, there is an interesting paradox related to that point.

Since the slowness and the cost of the certification processes lead manufacturers to extend the life of their products as long as possible, some weird situations appear: Modern planes, such as the Boeing-777, still leave the factories with processors that were running PCs discarded more than 20 years ago from homes, not places requiring high processing capacity.

Of course, the manufacturer of the processor (Intel) discarded that processor long ago, while Boeing must keep stock that could be considered IT archaeology, as long as the planes keep flying. Should the designers squeeze the bit in software design, while, at the same time, such outdated hardware is used?

At a time when the processor of a mobile phone can have about half the power of a fast PC processor and, of course, it is faster than a PC built during recent years, designing intelligible software could be a good investment in many fields.

> Peter Drucker, perhaps the best-known management guru of all time, gives an interesting example in his autobiography: During his work with a U.K. bank, he designed a complex investment strategy. His boss told him to show it to the least prominent of his coworkers. When Drucker took it almost as an offence, his boss told him that anything that was not understood by that guy should not be expected to be understood by the clients and, hence, it should be rejected.

The same principle should apply here: The top level in complexity design should be an easy place: It must be understood by the user. This does not need to be a universal rule since there are many activities that don't require it.

A temporary stop in the process or a mistake are not a big deal in many fields. Hence, tasks can be fragmented in the search for the most efficient way to perform them. However, if the stakes are higher and the solution to a problem must be found here and now, the top level of complexity design should be defined by an understanding level well beyond Windows knowledge.

The power of hardware allows designers to invest in intelligible software rather than nice interfaces that hide its real complexity. Therefore, the concept of being *transparent* to the user (a nice way to say "Don't bother about what's inside, it's not your business") should disappear, at least in high-stake activities where, additionally, a quick answer is required.

Of course, if we accept that deep knowledge about the system is required and Windows knowledge is not an acceptable option, we should probably carefully fine-tune the recruiting and training processes, to be sure that the intelligibility ceiling is not lower than desired. The idea of building intelligible systems leads to a question: For whom must they be intelligible? That question, in environments that require a fast and correct answer, has only one answer: the operator. So, an effort is required in both sides: Design of the system and profiles of the operators.

The Lack of Training Issue

Once more, lack of training, commonly used as an explanation in many major events, is hardly an acceptable explanation. When someone is performing a task and the training is not the adequate, we can ask why that person was performing that task. Any possible answer is going to lead us, at least, to organizational issues, not individual ones.

However, the problem is usually deeper: Many design problems can be disguised as "lack of training" and, in that way, point to the individual and, perhaps, to the wrong organization. That happens especially when an event is attributed to lack of training afterward.

Lack of training is a legitimate claim – that will never avoid the organizational side – when it comes before an event as a kind of warning. Once the event has happened, it becomes an easy label to try to remove any responsibility. Furthermore, beyond useless, this qualification leads to a patch-based training design.

Instead of looking for internal consistency, the last event qualified with this comfortable stamp leads to the inclusion of a patch to avoid that specific event. It does not help with the next one, which could be again assigned to lack of training and generate a new patch.

Of course, having someone overqualified for a job is expensive, but there are different ways to avoid it: If a task can be fragmented into little pieces and, depending on the skills required, different people can perform different parts, we will end up with a purely industrial approach like chain production. However, this process, which is easy to justify from the efficiency point of view, has some drawbacks that are not very visible.

These drawbacks appear in two different but related ways: flexibility and safety. A knowledge limited to specific tasks, whose links with different systems or places – sequences or relevance – in a whole process are ignored, makes changes of assignments difficult and, still worse, it opens the door to the wrong actions – or to inactions – without the knowledge of the operator.

Clearly, an organization designed with the parameters of a commando team is more expensive than an organization with strictly defined tasks and training related to them. However, that organization can answer to a wider range of contingencies than that which is strictly defined.

Furthermore, many major events happen because someone did not know the full impact of the action performed.

- The American Airlines 191 accident was linked to a maintenance practice in the removal of the engine from the wing. No one suspected that a poorly positioned lift cart could lead to a major accident.
- Spanair JKK-5022 happened after the removal of a piece that was in two different circuits. One of them – the circuit intended to

deactivate – was related to a heater for freezing conditions, while the other was related to the configuration alert, which was mistakenly disconnected.

- Aloha-243: The person checking the bolts on the fuselage never received a notice about the criticality of the task.
- G-KMAM: A maintenance team confused flaps and ailerons configurations, switching them.

Of course, these cases and many others could have been avoided with broader knowledge about the tasks to perform and where these tasks were to be carried out.

At the same time, one could say that the real problem is related to compliance, and that is true, as far as procedures are perfect and they deal with every possible contingency while performing the related task.

Since this ideal situation is quite far from reality, the concept of *exceptional violation* has become a reality, unthinkable in a perfect world. By the same token, concepts such as *malicious compliance* or *work-to-rule* would be perfect nonsense.

Knowledge and Performance

Knowledge and performance are very different but still related concepts that, for the sake of efficiency, have been separated further and further.

The old way to train people meant acquiring knowledge; after that, performance should appear. Since knowledge requires time to be acquired, before openly questioning that principle, there is a question regarding efficiency: What is the minimum amount of knowledge required to perform a task at the necessary quality level?

Unfortunately, the answer to that question is not as straightforward as it might seem. Theoretically, a fully detailed procedure, whether it is performed by people or by a machine, could lead to the right performance with the minimum knowledge – that is, the old approach in the Industrial Revolution or the more refined one developed by Frederick Taylor (1914).

The situation becomes more complex when new contingencies appear, leading to scenarios that are not included in the procedure. Then, the person or the machine would become blocked and the only way out is through knowledge.

The next question is about the frequency, relevance and urgency of situations that cannot be managed by a procedure or an algorithm.

A potential approach could be to define a value as a threshold above which the situation should be managed by people fully aware of the situation. Graphically, it could be represented by a table with two axes (potential

Potential Impact

21	21	42	63	105	168	273
13	13	26	39	65	104	169
8	8	16	24	40	64	104
5	5	10	15	25	40	65
3	3	6	9	15	24	39
2	2	4	6	10	16	26
1	1	2	3	5	8	13
1	1	2	3	5	8	13

Frequency

FIGURE 7.2
Fibonacci sequence used as a risk scale.

impact and frequency) whose values could grow according to a Fibonacci sequence (Figure 7.2).

A risk assessment could define the starting value in the table. After that, the appearance of new events would make the value grow. Once the value is above the threshold, full knowledge and awareness would be required.

Different units of frequency or different scales can be tried, but it would be a good principle to define where the *if-then* approach, based on sensors, effectors and a fixed algorithm linking them, would not be enough to manage the situation.

The table is trying to reflect that a failure whose expected outcome is trivial would not be a big deal, unless its frequency would be disturbing for the whole process or could lead to secondary effects by itself or combined with another failure.

If the outcome of a failure could be serious but there is enough time to manage it, the potential impact decreases, which would lead to a similar situation.

In both cases, unless the failure is frequent and, as such, disturbing to the process, a degraded situation could be accepted, if only temporarily. That's why planes are allowed to fly with minor technical problems, and the same rationale can be applied to training issues.

At the end of the day, this could be a common way to make a risk assessment and decide on an acceptable practice. However, there are cases where there can be no tolerance of lack of knowledge: situations with potentially serious outcomes and with little time to manage them – that is, situations with a potentially high impact.

Since the specific feature of Aviation is precisely that nobody can ask for a moment to look for someone else who knows better, the required knowledge together with the required authority must be on board the plane.

That does not mean reverting to a situation where, as a matter of fact, planes can perform tasks whose precision levels are far beyond human skills. For instance, tasks such as keeping the flight level in reduced vertical separation minimum (RVSM) space or performing a CAT III landing require full support from automatic systems.

However, these systems (or the pilot) can be misguided by a failure in the sensors or actuators or by their own internal processes.

The case of National Airlines 27, which took place in 1973 in Albuquerque, New Mexico, shows to what extent the pilot and flight engineer had only operational knowledge about how the autopilot worked, and this is far from being unique, of course. Many other cases have already been mentioned where the wrong information or the wrong automatic processes have produced major events.

These kind of events show why operational knowledge, or Windows knowledge as it is referred to here, instead of deep knowledge is the wrong solution in high-risk activities. It is especially unsuitable to environments where serious outcomes can appear in a short time and without the "pause" option.

What should be the solution? Expert technology to solve the nuisances coming from common technology? This solution and its effects can be better explained through a metaphor:

> The deepest levels in the deepest mines in the world cannot be reached with a single elevator, because the weight of the wire would break the wire. So, installing a stronger – and heavier – wire is not the solution.

In the same way, an algorithmic approach, designed to solve problems coming from an algorithmic approach, could be the wrong solution, especially if any solution is expected to deliver payback in terms of efficiency.

Senge gives us a first clue: When a system is near to capacity, the correct behavior is not to increase the effort to push the limits further but to work on them. Further improvement could be difficult, but the identification and removal of the limitations could be a more productive approach than insisting on the failed model.

Many people will deny that a model based on downgraded people and enhanced IT is a failed model, especially since it works fine in some environments. However, every resource has an optimal range of convenience, and trying to use it out of that range can lead to unexpected consequences.

Some designers try to avoid cases like the examples shown by building more complex algorithms that are less likely to be misguided by a sensor failure and with AI and Machine Learning as a part of the design. However, the first question before starting with that design is very simple:

Where is the payback for that development? If the project must be funded, and it is hard to imagine a project not requiring it, the project will be sold as a safety improvement, but at the same time it will bring an important efficiency improvement. Otherwise, the project will die.

Many examples have been shown displaying that path: Better altimetry leading to shorter distances between flying planes, better landing support leading to operations in airports that would otherwise be closed, better navigation resources leading to the disappearance of the navigator, better automation leading to the disappearance of the flight engineer and so on. More examples could be found, but all of them have a common feature: There is a trade-off.

Unless there is a major and alarming event that requires an immediate and specific change – cases like the cargo door of DC10 or confusions from multimodal instruments like the Mount St. Odile accident – every major change is funded only if it shows that, together with keeping or increasing the safety level, it can increase efficiency.

Since efficiency increases often come from having fewer and less trained people, raising the training issue is, at the same time, necessary and opposite to the present trend.

The last developments are pointing to have single pilots in big planes – cargo planes in the first phase – supported by more evolved automation and information systems and, if required, controllable from the ground. In other words, the *scientific organization of work* is still alive, even in places where it should not be.

By doing so, using the term coined by Perrow, organizations become more and more *tightly coupled* and, hence, prone to snowball effects coming from minor unforeseen failures. Perrow's tightly coupled concept can be read in different terms. A suggestion to read it in the right way would be as follows:

Errors, in efficient organizations, are also efficient. They spread their effects by using the same channels that the organization has created to operate efficiently. So, asking for efficiency in safety improvements could be the best way to have fewer but less manageable problems.

Human Value and the Conditions in Which to Use It

The key questions to be answered at this point are about specific human features, whether they are worth keeping, and why, when and where.

The development path has been openly opposite. Actually, efficiency improvements often come at the cost of people. So, we should question whether this is the right path, at least regarding safety-critical activities, and if not, what the alternative path should be.

If the most valuable human role in safety-critical activities must be defined, it would be the breaker. Situations that can enter into a self-reinforcing loop or suffer the snowball effect can be broken by human intervention. However, the effectiveness of this role is not guaranteed simply with human presence.

As Edgar Morin said, "The machine-machine is always superior to the man-machine." Hence, if people are expected to behave as machines, the inconveniences of an imperfect machine should be expected and, at the same time, the advantages of humans would be lost.

Paradoxically, this situation – putting people into environments that are better suited for machines – is used to justify why human presence should decrease and, as much as possible, the place of people should be taken by increasingly complex systems.

What is the right option? At the very least, in safety-critical environments like Aviation, the options should not be made invisible nor impeded by automatic processes. Performance improvement should not be an excuse not to meet this condition.

Jef Raskin, who was on the design team for the first Macintosh, clearly explains two Human Factors principles to be kept in the design of different objects, using as examples watches and car radios.

1. New functions can be admitted, but those that already exist should not be made more complicated than before.
2. Clean interfaces can hide complex functioning and the reverse; more buttons do not mean more complexity.

Both principles are so obvious and, at the same time, ignored.

In summary, there are many fields where the value added by human intervention is low while in others – especially in open environments, high-risk activities and time-constrained operations – people are required.

In the first case, people have been disappearing, and many jobs that were common some years ago have decreased in number or fully disappeared.

The second case is more complex: The idea seems to be to use people as an emergency resource, and perhaps the idea is sound, especially since the precision requirements in some cases are far beyond the reach of human beings.

However, even if this idea is accepted, some conditions must be met, especially regarding what information to use and when, how it is organized, what tasks must be performed, what is the value of these tasks, and whether they are situation awareness enablers.

Designing user-friendly interfaces and meeting the physical ergonomic requirements is necessary but it is not enough, even if the assigned role is as limited as being an emergency resource.

An emergency resource, to work properly, must be present during normal operation – that is, while the emergency is developing. In that way, there is a chance to stop the emergency and, furthermore, clues are acquired about its origin.

Notes

1. Louis Matsakis, Researchers fooled a Google AI into thinking a rifle was a helicopter, *Wired*, December 20, 2017. https://www-wired-com.cdn.ampproject.org/c/s/www.wired.com/story/researcher-fooled-a-google-ai-into-thinking-a-rifle-was-a-helicopter/amp.
2. After the U232 case, the system on board the plane that would take the place of the DC10 (the MD11) was designed to manage similar events.

Bibliography

Air Accidents Investigation Branch (1995). Report 2/95: Report on the incident to Airbus A320-212, G-KMAM London Gatwick Airport on 26 August 1993.

Air Accidents Investigation Branch (2014). Formal report AAR 1/2010: Report on the accident to Boeing 777-236ER, G-YMMM, at London Heathrow Airport on 17 January 2008.

Australian Transport Safety Bureau (2013). AO-2010-089 Final investigation: In-flight uncontained engine failure Airbus A380-842, VH-OQA Overhead Batam Island, Indonesia, 4 November 2010.

Asimov, I. (1989). *Azazel*. Barcelona: Plaza y Janés.

Asimov, I. (1951). *I, Robot* (Vol. 1). New York: Gnome Press.

Aviation Safety Network (2007). Turkish Airlines Flight 981. Accident report, accessed November, 23. https://aviation-safety.net/database/record.php?id=19740303-1

Bainbridge, L. (1983). Ironies of automation. In: *Analysis, Design and Evaluation of Man–Machine Systems 1982* (pp. 129–135).Oxford: Pergamon.

Batchelor, T. (2018). Single-pilot passenger planes could soon take to the skies, says Boeing. *Independent*, August 26. https://www.independent.co.uk/travel/news-and-advice/single-pilot-plane-boeing-autonomous-jet-technology-cockpit-a8506301.html.

Beck, U. (2002). *La sociedad del riesgo globalRisk Society*. Siglo XXI, Madrid.

Bennett, K. B., & Flach, J. M. (2011). *Display and Interface Design: Subtle Science, Exact Art*. Boca Raton, FL: CRC Press.

Boudette, N. (2018). Fatal Tesla crash raises new questions about autopilot system. New York Times, March 31. https://www.nytimes.com/2018/03/31/business/tesla-crash-autopilot-musk.html.

Brooks, R. A. (2003). *Flesh and Machines: How Robots Will Change Us*. Vintage.

Bureau d'Enquêtes et d'Analyses pour la sécurité de l'aviation civile (1993). Official report into the accident on 20 January 1992 near Mont Sainte-Odile (Bas-Rhin) of the Airbus A320 registered F-GGED operated by Air Inter.

Bureau d'Enquêtes et d'Analyses pour la sécurité de l'aviation civile (2010). Report on the accident on 27 November 2008 off the coast of Canet-Plage (66) to the Airbus A320-232 registered D-AXLA operated by XL Airways German.

Bureau d'Enquêtes et d'Analyses pour la sécurité de l'aviation civile (2012). Final report on the accident on 1st June 2009 to the Airbus A330-203 registered F-GZCP operated by Air France Flight AF 447 Rio de Janeiro–Paris.

Business insider (2018). If you have a Tesla and use autopilot, please keep your hands on the steering wheel. April 2. https://www.businessinsider.com/tesla-autopilot-drivers-keep-hands-on-steering-wheel-2018-4?IR=T.

Chess Network (2017). Google's self-learning AI AlphaZero masters chess in 4 hours. YouTube, December 7. https://youtu.be/0g9SlVdv1PY.

Coe, T., Mathisen, T., Moler, C., & Pratt, V. (1995). Computational aspects of the Pentium affair. *IEEE Computational Science and Engineering*, 2(1), 18–30.

Comisión de Investigación de Accidentes e Incidentes de Aviación Civil (2011). A-032/2008 Accidente ocurrido a la aeronave McDonnell Douglas DC-9-82 (MD-82), matrícula EC-HFP, operada por la compañía Spanair, en el aeropuerto de Barajas el 20 de agosto de 2008.

Copeland, B. J. (1998). Turing's O-machines, Searle, Penrose and the brain. *Analysis*, 58(2), 128–138.
Dekker, S. (2017). *The Field Guide to Understanding Human Error.* Boca Raton, FL: CRC Press.
Dennett, D. C. (2008). *Kinds of Minds: Toward an Understanding of Consciousness.* New York: Basic Books.
DoD. (2012). *Department of Defense Design Criteria Standard: Human Engineering (MIL-STD-1472G).* Washington, DC: Department of Defense.
Dismukes, R. K., Berman, B. A., & Loukopoulos, L. (2007). *The Limits of Expertise: Rethinking Pilot Error and the Causes of Airline Accidents.* Aldershot, UK: Ashgate.
Dismukes, R. K., Kochan, J. A., & Goldsmith, T. E. (2018). Flight crew errors in challenging and stressful situations. *Aviation Psychology and Applied Human Factors*, 8(1), 35–46.
Dörner, D., & Kimber, R. (1996). *The Logic of Failure: Recognizing and Avoiding Error in Complex Situations* (Vol. 1). New York: Basic Books.
Dreyfus, H., Dreyfus, S. E.,, T. (2000). *Mind over Machine.* New York: Simon and Schuster
Dreyfus, H. L. (1992). *What Computers Still Can't Do: A Critique of Artificial Reason.* Cambridge, MA: MIT Press.
Drucker, P. (1978). *Adventures of a Bystander.* New York: Harper and Row.
EASA, CS25 (2018). Certification specifications for large aeroplanes: Amendment 22. https://www.easa.europa.eu/certification-specifications/cs-25-large-aeroplanes.
Endsley, M. R. (2015). Final reflections: Situation awareness models and measures. *Journal of Cognitive Engineering and Decision Making*, 9(1), 101–111.
Fischhoff, B., Lichtenstein, S., Slovic, P., Derby, S. L., & Keeney, R. (1983). *Acceptable Risk.* Cambridge: Cambridge University Press.
Fleming, E., & Pritchett, A. (2016). SRK as a framework for the development of training for effective interaction with multi-level automation. *Cognition, Technology and Work*, 18(3), 511–528.
Gallo, R. C., & Montagnier, L. (2003). The discovery of HIV as the cause of AIDS. *The New England Journal of Medicine*, 349(24), 2283–2285.
Gibbs, S. (2017). AlphaZero AI beats champion chess program after teaching itself in four hours. *Guardian*, December 7. https://www.theguardian.com/technology/2017/dec/07/alphazerogoogle-deepmind-ai-beats-champion-program-teaching-itself-to-play-four-hours.
Gigerenzer, G., & Todd, P. M. (1999). *Simple Heuristics that Make Us Smart.* Oxford: Oxford University Press.
Hawkins, J., & Blakeslee, S. (2004). *On Intelligence: How a New Understanding of the Brain Will Lead to the Creation of Truly Intelligent Machines.* New York: Owl Books.
Kahneman, D., & (2011). *Thinking, Fast and Slow* (Vol. 1). New York: Farrar, Straus and Giroux.
Klein, G. (2013). *Seeing What Others Don't.* New York: Public Affairs.
Klein, G. A. (1993). *A Recognition-Primed Decision (RPD) Model of Rapid Decision Making* (pp. 138–147). New York: Ablex.
Kozaczuk, W. (1984). Enigma: How the German Machine Cipher Was Broken, and How It Was Read by the Allies in World War Two. *Foreign Intelligence Book Series.* Lanham, MD: University Publications of America.
Leveson, N. (2011). *Engineering a Safer World: Systems Thinking Applied to Safety.* Cambridge, MA: MIT Press.

Leveson, N. G., & (1995). *Safeware: System Safety and Computers* (Vol. 680). Reading, MA: Addison-Wesley.
Liskowsky, D. R., & Seitz, W. W. (2010). *Human Integration Design Handbook*. Washington, DC: National Aeronautics and Space Administration (NASA), pp. 657–671.
Luhmann, N. (1993). *Risk: A Sociological Theory*. New York: A. de Gruyter.
Maurino, D. E., Reason, J., Johnston, N., & Lee, R. B. (1995). *Beyond Aviation Human Factors: Safety in High Technology Systems*. Aldershot, UK: Ashgate.
Morin, E., & Pakman, M. (1994). *Introducción al pensamiento complejo*. Barcelona: Gedisa.
National Transportation Safety Board (1973). NTSB-AAR-73-19 Uncontained engine failure, National Airlines, Inc., DC-10-10, N60NA, near Albuquerque, New Mexico, November 3, 1973.
National Transportation Safety Board (1979). NTSB/AAR-79/17 American Airlines DC10-10, N110AA Chicago, O'Hare International Airport, Chicago-Illinois, May 25, 1979.
National Transportation Safety Board (1990). NTSB/AAR-90/06 United Airlines Flight 232. McDonnell Douglas DC-I0-10 Sioux Gateway Airport Sioux City, Iowa, July 19, 1989.
National Transportation Safety Board (2000). NTSB/AAB-00/01 Accident no.: DCA00MA005; Operator or flight number: Sunjet Aviation; Aircraft and registration: Learjet Model 35, N47BA, Aberdeen, South Dakota, October 25, 1999.
Parasuraman, R., Sheridan, T. B., & Wickens, C. D. (2008). Situation awareness, mental workload, and trust in automation: Viable, empirically supported cognitive engineering constructs. *Journal of Cognitive Engineering and Decision Making*, 2(2), 140–160.
Penrose, R., N. D. (1990). *The Emperor's New Mind: Concerning Computers, Minds, and the Laws of Physics*. Oxford: Oxford Landmark Science.
Perrow, C. (2011). *Normal Accidents: Living with High Risk Technologies-Updated Edition*. Princeton, NJ: Princeton University Press.
Petroski, H. (1985). *To Engineer Is Human: The Role of Failure in Successful Design*. New York: Vintage.
Preston, J., & Bishop, M. J. (2002). *Views into the Chinese Room: New Essays on Searle and Artificial Intelligence*. Oxford: Oxford University Press.
Raskin, J. (2000). *The Humane Interface: New Directions for Designing Interactive Systems*. Reading, MA: Addison-Wesley Professional.
Rasmussen, J. (1986). *Information Processing and Human–Machine Interaction: An Approach to Cognitive Engineering, North-Holland Series in System Science and Engineering, 12*. New York: North-Holland.
Rasmussen, J., & Vicente, K. J. (1989). Coping with human errors through system design: Implications for ecological interface design. *International Journal of Man–Machine Studies*, 31(5), 517–534.
Reason, J. (2017). *The Human Contribution: Unsafe Acts, Accidents and Heroic Recoveries*. Boca Raton, FL: CRC Press.
Risukhin, V. (2001). *Controlling Pilot Error: Automation*. New York: McGraw-Hill.
Rudolph, J., Hatakenaka, S., & Carroll, J. S. (2002). Organizational Learning from Experience in High-Hazard Industries: Problem Investigation as Off-Line Reflective Practice. MIT Sloan School of Management Working Paper 4359-02.

SAE (2003). Human Factor considerations in the design of multifunction display systems for civil aircraft. ARP5364. https://www.sae.org/standards/content/arp5364/
Sánchez-Alarcos Ballesteros, J. (2007). *Improving Air Safety through Organizational Learning: Consequences of a Technology-Led Model*. Aldershot, UK: Ashgate.
Secretariat d'Etat aux Transports (1976). Rapport Final de la Commission d'Enquête sur l'accident de l'avion D.C.·10 TC:JAV des Turkish Airlines survenu à Ermenonville, 14e, 3 mars 1914.
Senate of the United States (2018). Calendar No. 401 115th Congress 2D Session H. R. 4 in the Senate of the United States; May 7, 2018 received, read the first time; May 8, 2018 read the second time and placed on the calendar.
Senge, P. M. (2010). *The Fifth Discipline, the Art and Practice of the Learning Organization*. London: Cornerstone Digital.
Stanton, N. A., Harris, D., Salmon, P. M., Demagalski, J., Marshall, A., Waldmann, T., et al. (2010). Predicting design-induced error in the cockpit. *Journal of Aeronautics, Astronautics and Aviation*, 42(1), 1–10.
Taylor, F. W. (1914). *The Principles of Scientific Management*. Harper.New York
Tesla (2018). Full self-driving hardware on all cars. https://www.tesla.com/autopilot.
Tognotti, E. (2011). The dawn of medical microbiology: Germ hunters and the discovery of the cause of cholera. *Journal of Medical Microbiology*, 60(4), 555–558.
Turing, A. M. (1950). Computing machinery and intelligence. *Mind* 49 pp. 433–460.
Transportation Safety Board Canada (1985). Final report of the board of inquiry investigating the circumstances of an accident involving the Air Canada Boeing 767 aircraft C-GAUN that effected an emergency landing at Gimli, Manitoba on the 23rd day of July, 1983; Commissioner, the Honourable Mr. Justice George H. Lockwood April 1985.
Vicente, K. J. (1999). *Cognitive Work Analysis: Toward Safe, Productive, and Healthy Computer-Based Work*. Boca Raton, FL: CRC Press.
Vicente, K. J. (2002). Ecological interface design: Progress and challenges. *Human Factors*, 44(1), 62–78.
Vicente, K. J., & Rasmussen, J. (1992). Ecological interface design: Theoretical foundations. *IEEE Transactions on Systems, Man, and Cybernetics*, 22(4), 589–606.
Wikipedia (2019). Pentium FDIV bug. https://en.wikipedia.org/wiki/Pentium_FDIV_bug.
Winograd, T., Flores, F., & Flores, F. F. (1986). *Understanding Computers and Cognition: A New Foundation for Design*. New York: Addison-Wesley.
Wood, S. (2004). *Flight Crew Reliance on Automation*. CAA Paper 10.

8

Conclusions

The different cases shown in the preceding chapters could be considered samples of a trend that started long ago – decreasing, as much as possible, the role of Human Factors. However, since this fact not only affects Aviation, one could legitimately ask whether there is a real problem, whether this problem is general or whether, in any way, that single fact affects Aviation especially.

Statistical information refers to an activity that learns from mistakes over time, and the decrease in the pace of improvements could be explained not only by an exhausted learning model but by the physical impossibility of going any further.

Nowadays, while thousands of flights are performed every day, a single major event is enough to show a peak in the graphical information. Hence, despite the major differences between advanced and developing countries, Aviation can qualify as a fairly safe business.

Furthermore, distressing statements from the White House in 1998 forecast weekly accidents by 2015. Of course, forecasting the past is easier than forecasting the future but, in any case, it seems that the White House Commission was unnecessarily alarmist. Then again, is something wrong?

In spite of the preceding facts, the answer is affirmative. Aviation has shared a model of evolution with many other activities, and by doing so, specific features of high-risk activities and, specifically, Aviation have not been kept in mind.

More than a century ago, the Industrial Revolution started a movement that continues today: standardizing human behavior, trying to make people easy to replace, and getting the most out of people as ordinary as possible and with the least possible amount of training.

That general principle seemed to be valid for mechanical activities requiring few qualifications. Of course, once technology was able to provide an alternative resource, it would be used in preference to people. That can be criticized from the social point of view, but from an efficiency point of view it worked.

World War II brought that general principle to Aviation, due to the urgent need for flight crews that could be trained as quickly as possible. Thousands of planes (bombers, fighters, transporters, etc.) had to be flown and the time available to find people to fly them was short. Then, efficiency, as classically defined, rose to the highest possible levels in order to do as much as possible with as few resources as possible.

After the war, a new factor arose: the evolution of information technology (IT). Once IT became important, the principles of the Industrial Revolution were extended to more specialized tasks, including highly specialized jobs in the Aviation environment.

While this development was taking place, discourses about the value of Human Factors were very common. However, these discourses were hiding some facts: In the same way that the main focus changed from people to systems/machines, there was another movement regarding the relevance of different people. The people operating the systems were losing relevance in favor of the people designing those systems. For obvious reasons, the number of designers is always lower than the number of operators; hence, the system gains efficiency by behaving that way.

This has been the development path. People performing the operational tasks have been fed with procedural knowledge. The increasing complexity of the environment led to a trade-off between deep knowledge and blind procedural compliance. Deep knowledge – limited by very narrow specialties – became the privilege of design teams.

These human–machine and design–operation trade-offs were far from being free. The number of accidents decreased, but at the same time the factors behind them changed.

An increase in technical reliability led to a decrease in technical faults. In some cases, that increase in technical reliability led to a decrease in redundancy (e.g. ETOPS), statistically supported by that improved reliability.

At the same time, a new kind of accident emerged where confusion would be the main driver. Different scenarios have appeared that are difficult to anticipate in the current training processes, where faulty sensors trigger automatic processes and make both assessment and control difficult (e.g. XL888T, AF447 or Lion Air 610). Furthermore, there is no standard behavior: Different planes can share identical systems, making it easier to miss a mistaken parameter, and different automation models can be found in different manufacturers (Asiana 214), leading to confusion if the crew are unfamiliar with the automation design.

This new fashion is not entirely new, and accidents with these features were seen in the last few decades of the previous century (e.g. AeroPeru 603). However, since the first few years of the present century, new accidents became the main source of major events in Commercial Aviation.

Operations officers are supposed to behave like firemen, willing to intervene in an emergency, but to do so, some basics about how to remain in the loop should be respected, and they are not.

The consequence of the evolution up to the present stage has been superb performance in familiar situations. At the same time, that performance has decreased in unfamiliar situations or in emergencies, where confusion has been, very often, the main element.

The most interesting side of it is the unsound learning acquired from these facts: If people are there basically to manage emergencies and they fail time and again, systems should be designed as "idiot-proof", at least while it's not feasible to do away with them. That's precisely the wrong lesson to learn.

Idiot-proof designs are usually based on graphically pleasing interfaces, filled with metaphors about how everything works (*Windows knowledge*) and with a complex structure of warnings and automatic processes that, in some cases, are even willing to overrule the human supposedly in charge.

This trend has added complexity to the whole system. People have become still more prone to confusion and its resulting events. Therefore, since the idiot-proof way did not work, the alternative – getting rid of people – has begun. IT, within the AI or Machine Learning models, is again supposed to be the workhorse to assist the main Aviation players in this task.

That could be a summary of the evolution and the present stage. From a technical point of view, there is nothing to be worried about as far as numbers show that air safety keeps increasing, albeit at a slower pace.

Even so, a remark must be made: Safety figures have been improving not only because of technology improvements. The human side has been there to stop sequences that, by themselves, could lead to accidents. Eliminating or decreasing the human role would also eliminate this firewall. Therefore, in this process, safety figures could suffer, showing that evolution is heading the wrong way.

Commonly used data about increasing technological reliability and, therefore, an increasing number of human failures (in percentage terms) have been incorrectly read as an invitation to increase technology and decrease humanity.

Leaving aside the single fact that a reliable device – or a whole system – can lead to human error because of poor design or organizational variables, people have always been there to solve bizarre situations generated by a reliable technology working out of the foreseen context.

However, beyond a serious warning about the possibility of worsening the present figures by following the present path, this complacent attitude related to the role of Human Factors and their connection to air safety is missing two major points:

1. People are willing to accept dangers as a part of life, but these same people will reject *manufactured dangers* – that is, situations that come from a risk analysis, accepted by regulators or the high-ranking managers of operators, but not by the users.

   As the aforementioned Tesla case (the statement by the company and the reaction of the public) has shown, an explanation pointing out that, although the victim of an accident is dead, the chances of being killed in a human-oriented environment are still higher does not work. Furthermore, it would be considered outrageous.

2. Nobody can control an information tsunami. The misbehavior of a company will spread at light speed without any possible brakes or

> control. The fall of McDonnell Douglas took years after the TK981 case, but in a Web 2.0 environment the outcome could be much faster. An undesired "trending topic" can drive anyone out of business at a much faster speed than in the past.

So, there is good reason to be worried, and it is not only related to potential changes on the technical side or in the statistical information. It is in an eventual uncontained reaction by the public to a hidden *manufactured uncertainty*. Some decisions that are already a part of common operations are fully unknown by the public. Some others could follow the same path in the future, and if so, a potential rejection by the public would be much more serious than in the past.

The "superior rationality of experts" can crash against the *disaster threshold* used by common people as one of the main decision criteria. That is a serious risk in the present world, but there is good news: It can be prevented by learning the right lessons.

The wrong lessons have led to the manufactured incompetence of Human Factors by giving up everything beyond anthropometrics and physical ergonomics. Human Factors, as a discipline, has often been – sometimes, explicitly – centered in Error Avoidance but, as such, it can also spoil the options for a sound answer when things become difficult.

In some ways, the safest way to avoid manual errors is to "handcuff" the operator but, at the same time, that operator will be unable to solve a problem when required. Handcuffs can be physical, such as programming a plane to disobey the pilot because someone has decided that the sensors of the plane would know better, or cognitive, providing operating *know-how* instead of *know-why*. Accidents coming from these origins, widely shown here, are usually assigned to human error, be it lack of training or another variety, but the main fact remains: The human error was triggered by a minor failure on the technology side, whose effects were multiplied by inadequate design.

Additionally, many apparently trivial human interventions can stop the sequence leading to an emergency or a major event and, again, this sequence can be begun by one of the many situations that technology and procedures did not foresee.

Only the most prominent of these situations (e.g. QF32, US1549 and some others) are known, but many other apparently minor interventions are simply ignored or considered a part of the routine. Hence, the data on the real human contribution are biased in favor of technology.

Statistics about human error can tell the truth but not the whole truth; that is, when someone says how many events come from human error, there is an unknown counterpart that should include how many events are avoided because of human intervention. Regretfully, the information about this is very scarce. Even reporting systems have databases about near-misses but not about actions that, despite breaking a chain that could have lead to the event, could be considered ordinary by the people performing them.

In this process of downgrading the human element, cognitive ergonomics is commonly forgotten and, hence, when people are called to intervene, they miss the most basic pieces or, worse still, the pieces that they have are misguiding.

Nice interfaces and blind procedure compliance are a poor substitute for functional knowledge, where the human operator knows the functional relationship between the different parts of the system, making the assessment and the intervention possible.

Therefore, the next step is not in improving IT by using available resources or telling of the wonders of others to come. Neither is it a matter of increasing efforts toward the elimination of human error, as a prior step in the elimination of the human itself.

The solution requires creating a human-friendly environment that can supply the elements required to find a solution. It must be noted once again that Aviation has a very specific feature:

A flying plane is an isolated piece of an organization. Although high-ranking managers and subject experts can remain on the ground, the available resources for a quick answer in an unplanned situation are basically those on board the plane. So, authority and knowledge must also be on board and at a level high enough to manage the situation.

That's why an organizational model that favors systems at the cost of people, so common in many activities, deserves to be criticized when used precisely in Aviation.

Decreasing the quality of recruitment and training and designing planes to be managed by very average people was justified in a situation like World War II. Nowadays, the only justification is economic – that is, bringing the mass-production model to an environment where it does not fit because of its specific features.

If Aviation's main players want to learn the right lessons, a change of path is required: Technology must be designed to make its functionality and the relationships between the different parts visible – that is, understandable by the user.

At the same time, the emphasis on error avoidance should be changed to an emphasis on making insight possible in abnormal situations. That emphasis will change the tasks to be performed, which should be evaluated in terms of their role as situation awareness enablers. Therefore, the training processes and the recruiting processes should be adapted to different skills and knowledge requirements.

This change will not eliminate all the accidents, although it could at least eliminate those coming from confusion. Actually, nothing can eliminate all accidents, whether it is based on people, technology, procedures or organizational models. However, within this environment, if a new accident happens, nobody will be able to say that every person responsible at every level did not make the best effort to avoid it.

That's the challenge. Admitting it and confronting it is up to the management of the different Aviation players.

Bibliography

Abbott, K., Slotte, S. M., & Stimson, D. K. (1996). The interfaces between flightcrews and modern flight deck systems. FAA HF Team report.

Ahlstrom, V., & Longo, K. (2003). Human Factors Design Standard (HF-STD-001). Atlantic City International Airport, NJ: Federal Aviation Administration William J. Hughes Technical Center.

Air Accidents Investigation Branch (1995). Report 2/95: Report on the incident to Airbus A320-212, G-KMAM London Gatwick Airport on 26 August 1993.

Air Accidents Investigation Branch (2014). Formal report AAR 1/2010: Report on the accident to Boeing 777-236ER, G-YMMM, at London Heathrow Airport on 17 January 2008.

Australian Transport Safety Bureau (2013). AO-2010-089 Final Investigation: In-flight uncontained engine failure Airbus A380-842, VH-OQA overhead Batam Island, Indonesia, 4 November 2010.

Air Force, U.S. (2008). *Air Force Human Systems Integration Handbook*. Dayton, OH: Air Force 711 Human Performance Wing, Directorate of Human Performance Integration, Human Performance Optimization Division.

Asimov, I. (1988). *Azazel*. Doubleday.

Asimov, I. (2004). *I, Robot* (Vol. 1). Spectra.

ASRS: Aviation Safety Reporting System (2019). ASRS Database Online. https://asrs.arc.nasa.gov/search/database.html.

Bainbridge, L. (1983). Ironies of automation. In: *Analysis, Design and Evaluation of Man–Machine Systems 1982* (pp. 129–135).Oxford: Pergamon.

Beaty, D. (1991). *The Naked Pilot: The Human Factor in Aircraft Accidents*. Airlife.

Beck, U. (2002). *La sociedad del riesgo global*. Madrid: Siglo XXI.

Bennett, K. B., & Flach, J. M. (2011). *Display and Interface Design: Subtle Science, Exact Art*. Boca Raton, FL: CRC Press.

Booher, H. R. (2012). *MANPRINT: An Approach to Systems Integration*. Springer Science & Business Media.

Brooks, R. A. (2003). *Flesh and Machines: How Robots Will Change Us*. New York: Pantheon Books.

Bureau d'Enquêtes et d'Analyses pour la sécurité de l'aviation civile (1993). Official report into the accident on 20 January 1992 near Mont Sainte-Odile (Bas-Rhin) of the Airbus A320 registered F-GGED operated by Air Inter.

Bureau d'Enquêtes et d'Analyses pour la sécurité de l'aviation civile (2010). Report on the accident on 27 November 2008 off the coast of Canet-Plage (66) to the Airbus A320-232 registered D-AXLA operated by XL Airways German.

Bureau d'Enquêtes et d'Analyses pour la sécurité de l'aviation civile (2012). Final report on the accident on 1st June 2009 to the Airbus A330-203 registered F-GZCP operated by Air France Flight AF 447 Rio de Janeiro–Paris.

Burkill, M., & Burkill, P. (2010). *Thirty Seconds to Impact*. Central Milton Keynes: AuthorHouse.

Business Insider (2018). "If you have a Tesla and use autopilot, please keep your hands on the steering wheel", April 2. https://www.businessinsider.com/tesla-autopilot-drivers-keep-hands-on-steering-wheel-2018-4?IR=T.

Comisión de Investigación de Accidentes e Incidentes de Aviación Civil (2011). A-032/2008: Accidente ocurrido a la aeronave McDonnell Douglas DC-9-82 (MD-82), matrícula EC-HFP, operada por la compañía Spanair, en el aeropuerto de Barajas el 20 de agosto de 2008.

Dekker, S. (2017). *The Field Guide to Understanding 'Human Error'*. Boca Raton, FL: CRC Press.

Dennett, D. C. (2008). *Kinds of Minds: Toward an Understanding of Consciousness*. New York: Basic Books.

Diaper, D., & Stanton, N. (Eds.) (2003). *The Handbook of Task Analysis for Human-Computer Interaction*. Boca Raton, FL: CRC Press.

Dismukes, R. K., Berman, B. A., & Loukopoulos, L. (2007). *The Limits of Expertise: Rethinking Pilot Error and the Causes of Airline Accidents*. Aldershot, UK: Ashgate.

DoD (2012a). Department of Defense design criteria standard: Human engineering (MIL-STD-1472G). Washington, DC: Department of Defense.

DoD (2012b). Mil-std-882e, Department of Defense standard practice system safety. Washington, DC: US Department of Defense.

Dörner, D., & Kimber, R. (1996). *The Logic of Failure: Recognizing and Avoiding Error in Complex Situations* (Vol. 1). New York: Basic Books.

Dreyfus, H., Dreyfus, S. E., & (1986). *Mind over Machine*. New York: Free Press.

Dreyfus, H. L. (1992). *What Computers Still Can't Do: A Critique of Artificial Reason*. Cambridge, MA: MIT Press.

EASA, CS25 (2018). Certification specifications for large aeroplanes: Amendment 22. https://www.easa.europa.eu/certification-specifications/cs-25-large-aeroplanes.

Edvinsson, L., & Malone, M. S. (1997). *Intellectual Capital: Realizing Your Company's True Value by Finding Its Hidden Brainpower*. New York: Harper Business.

Endsley, M. R. (2015). Final reflections: Situation awareness models and measures. *Journal of Cognitive Engineering and Decision Making*, 9(1), 101–111.

FAA (2016.) Human Factors design standard. FAA Human Factors Branch.

Fielder, J., & Birsch, D. (1992). *The DC-10 Case: A Study in Applied Ethics, Technology, and Society*. New York: State University of New York Press.

Fischhoff, B., Lichtenstein, S., Slovic, P., Derby, S. L., & Keeney, R. (1983). *Acceptable Risk*. Cambridge: Cambridge University Press.

Fleming, E., & Pritchett, A. (2016). SRK as a framework for the development of training for effective interaction with multi-level automation. *Cognition, Technology and Work*, 18(3), 511–528.

Gabinete de Prevençao e Investigaçao de Acidentes com Aeronaves (2001). Accident investigation final report: All engines-out landing due to fuel exhaustion; Air Transat Airbus A330-243 Marks C-GITS Lajes, Azores, Portugal, 24 August 2001.

Gigerenzer, G. (2007). *Gut Feelings: The Intelligence of the Unconscious*. New York: Penguin.

Gigerenzer, G., & Todd, P. M. (1999). *Simple Heuristics that Make Us Smart*. Oxford: Oxford University Press.

Hale, A. R., Wilpert, B., & Freitag, M. (Eds.) (1997). *After the Event: From Accident to Organisational Learning*. New York: Elsevier.

Harris, D. (2011). Rule fragmentation in the airworthiness regulations: A human factors perspective. *Aviation Psychology and Applied Human Factors*, 1(2), 75–86.

Hawkins, J., & Blakeslee, S. (2004). *On Intelligence: How a New Understanding of the Brain Will Lead to the Creation of Truly Intelligent Machines*. New York: Owl Books.

Hillis, W. D. (1999). *The Pattern on the Stone: The Simple Ideas that Make Computers Work*. New York: Basic Books.

Hubbard, D. W. (2009). *The Failure of Risk Management: Why It's Broken and How to Fix It*. New York: John Wiley.

ICAO (2018). Doc. 9859. *Safety Management Manual*, 4rd ed., ICAO Montreal.

Kahneman, D., & Egan, P. (2011). *Thinking, Fast and Slow* (Vol. 1). New York: Farrar, Straus and Giroux.

Klein, G. (2013). *Seeing What Others Don't*. New York: Public Affairs.

Klein, G. A. (1993). *A Recognition-Primed Decision (RPD) Model of Rapid Decision Making* (pp. 138–147). New York: Ablex.

Komite Nasional Keselamatan Transportasi Republic of Indonesia (2018). Preliminary KNKT.18.10.35.04 aircraft accident investigation report PT: Lion Mentari Airlines Boeing 737-8 (MAX); PK-LQP Tanjung Karawang, West Java Republic of Indonesia, 29 October 2018.

Leveson, N. (2011). *Engineering a Safer World: Systems Thinking Applied to Safety*. Cambridge, MA: MIT Press.

Leveson, N. G. (1995). *Safeware: System Safety and Computers* (Vol. 680). Reading, MA: Addison-Wesley.

Leveson, N. G. (2004). Role of software in spacecraft accidents. *Journal of Spacecraft and Rockets*, 41(4), 564–575.

Liskowsky, D. R., & Seitz, W. W. (2010). *Human Integration Design Handbook* (pp. 657–671). Washington, DC: NASA.

Luhmann, N. (1993). *Risk: A Sociological Theory*. New York: A. de Gruyter.

Mansikka, H., Harris, D., & Virtanen, K. (2017). An input–process–output model of pilot core competencies. *Aviation Psychology and Applied Human Factors7*, pp. 78–85. https://doi.org/10.1027/2192-0923/a000120. © 2017 Hogrefe Publishing.

Maturana, H. R., & Varela, F. J. (1987). *The Tree of Knowledge: The Biological Roots of Human Understanding*. New Science Library/Shambhala.

Maurino, D. E., Reason, J., Johnston, N., & Lee, R. B. (1995). *Beyond Aviation Human Factors: Safety in High Technology Systems*. Aldershot, UK: Ashgate..

Ministry of Transports and Communications, Accidents Investigation Board (1996). Accident of the Boeing 757-200 aircraft operated by Empresa de Transporte Aéreo del Perú S.A., 2 October 1996. https://skybrary.aero/bookshelf/books/1719.pdf.

National Research Council (1997). *Aviation Safety and Pilot Control: Understanding and Preventing Unfavorable Pilot–Vehicle Interactions*. Washington, DC: National Academies Press.

National Transportation Safety Board (1973a). NTSB/AAR-73/2 American Airlines McDonnell Douglas DC10-10, N103AA near Windsor, Ontario, Canada, June 12, 1972.

National Transportation Safety Board (1973b). NTSB-AAR-73-19 Uncontained engine failure, National Airlines, Inc., DC-10-10, N60NA, near Albuquerque, New Mexico, November 3, 1973.

National Transportation Safety Board (1979). NTSB/AAR-79/17 American Airlines DC10-10, N110AA Chicago, O'Hare International Airport, Chicago-Illinois, May 25, 1979.

National Transportation Safety Board (1990). NTSB/AAR-90/06 United Airlines Flight 232, McDonnell Douglas DC-I0-10 Sioux Gateway Airport Sioux City, Iowa July 19, 1989.
National Transportation Safety Board (2010). NTSB/AAR-10/03 Loss of thrust in both engines after encountering a flock of birds and subsequent ditching on the Hudson River US Airways Flight 1549 Airbus A320-214, N106US Weehawken, New Jersey, January 15, 2009.
National Transportation Safety Board (2014). NTSB/AAR-14/01 Descent below visual glidepath and impact with seawall Asiana Airlines Flight 214 Boeing 777-200ER, HL7742 San Francisco, California, July 6, 2013.
O'Hara, J. M., Higgins, J. C., Fleger, S. A., & Pieringer, P. A. (2012). Human Factors engineering program review model. Washington, DC: Nuclear Regulatory Commission, Office of Nuclear Regulatory Research, Division of Risk Analysis.
Parasuraman, R., Sheridan, T. B., & Wickens, C. D. (2008). Situation awareness, mental workload, and trust in automation: Viable, empirically supported cognitive engineering constructs. *Journal of Cognitive Engineering and Decision Making*, 2(2), 140–160.
Penrose, R., & 990). *The Emperor's New Mind: Concerning Computers, Minds, and the Laws of Physics*. Oxford: Oxford Landmark Science.
Perrow, C. (1972). *Complex Organizations: A Critical Essay* New York: McGraw-Hill.
Raskin, J. (2000). *The Humane Interface: New Directions for Designing Interactive Systems*. Reading, MA: Addison-Wesley Professional.
Rasmussen, J. (1986). *Information Processing and Human–Machine Interaction: An Approach to Cognitive Engineering*, North-Holland Series in System Science and Engineering, *12*. New York: North-Holland.
Rasmussen, J., & Vicente, K. J. (1989). Coping with human errors through system design: Implications for ecological interface design. *International Journal of Man-Machine Studies*, 31(5), 517–534.
Reason, J. (1990). *Human Error*. Cambridge: Cambridge University Press.
Reason, J. (1997). *Managing the Risks of Organizational Accidents*. Aldershot, UK: Ashgate.
Reason, J. (2017). *The Human Contribution: Unsafe Acts, Accidents and Heroic Recoveries*. Boca Raton, FL: CRC Press.
Risukhin, V. (2001). *Controlling Pilot Error: Automation*. NewYork: McGraw-Hill.
Rudolph, J., Hatakenaka, S., & Carroll, J. S. (2002). Organizational learning from experience in high-hazard industries: Problem investigation as off-line reflective practice. MIT Sloan School of Management Working Paper 4359-02.
SAE (2003). *Human Factor Considerations in the Design of Multifunction Display Systems for Civil Aircraft*. ARP5364 https://www.sae.org/standards/content/arp5364/.
Sagan, S. D. (1995). *The Limits of Safety: Organizations, Accidents, and Nuclear Weapons*. Princeton, NJ: Princeton University Press.
Sánchez-Alarcos Ballesteros, J. (2007). *Improving Air Safety through Organizational Learning: Consequences of a Technology-Led Model*. Aldershot, UK: Ashgate.
Secretariat d'Etat aux Transports (1976). Rapport final de la Commission d'Enquête sur l'accident de l'avion D.C. 10 TC: JAV des Turkish Airlines survenu à Ermenonville, 14e, 3 mars 1974.
Senate of the United States (2018). Calendar no. 401 115th Congress 2D Session H. R. 4 in the Senate of the United States; May 7, 2018 received, read the first time; May 8, 2018, read the second time and placed on the calendar.

Senge, P. M. (2010). *The Fifth Discipline, The Art and Practice of the Learning Organization*. London: Cornerstone Digital.
Sowell, T. (1980). *Knowledge and Decisions* (Vol. 10). New York: Basic Books.
Taylor, F. W. (1914). *The Principles of Scientific Management*. New York: Harper.
Tesla (2018). Full self-driving hardware on all cars. https://www.tesla.com/autopilot.
Transportation Safety Board Canada (1985). Final report of the Board of Inquiry investigating the circumstances of an accident involving the Air Canada Boeing 767 aircraft C-GAUN that effected an emergency landing at Gimli, Manitoba on the 23rd day of July, 1983; Commissioner, the Honourable Mr. Justice George H. Lockwood April 1985.
Vicente, K. J. (1999). *Cognitive Work Analysis: Toward Safe, Productive, and Healthy Computer-Based Work*. Boca Raton, FL: CRC Press.
Vicente, K. J. (2002). Ecological interface design: Progress and challenges. *Human Factors*, 44(1), 62–78.
Vicente, K. J., & Rasmussen, J. (1992). Ecological interface design: Theoretical foundations. *IEEE Transactions on Systems, Man, and Cybernetics*, 22(4), 589–606.
Weick, K. E., & Sutcliffe, K. M. (2011). *Managing the Unexpected: Resilient Performance in an Age of Uncertainty* (Vol. 8). San Francisco: John Wiley.
Wheeler, P. H. (2007). Aspects of automation mode confusion. Doctoral dissertation, MIT, Cambridge, MA.
White House Commission on Aviation Safety and Security (1997). Final report to President Clinton, February 12, 1997. https://fas.org/irp/threat/212fin~1.html.
Winograd, T., Flores, F., & Flores, F. F. (1986). *Understanding Computers and Cognition: A New Foundation for Design*. New York: Addison-Wesley.
Wood, S. (2004). Flight crew reliance on automation. CAA Paper 10.

Index

C

D

E

F

G

H

For Product Safety Concerns and Information please contact our
EU representative GPSR@taylorandfrancis.com Taylor & Francis
Verlag GmbH, Kaufingerstraße 24, 80331 München, Germany